TEST YOUR MIND

For Completing Ordinary Level Mathematics

Second Edition

OMAR RAMADHAN KOMBO
Biomedical Engineer

authorHOUSE®

AuthorHouse™ UK
1663 Liberty Drive
Bloomington, IN 47403 USA
www.authorhouse.co.uk
Phone: 0800.197.4150

© 2018 Omar R Kombo. All rights reserved.

No part of this book may be reproduced, stored in a retrieval system, or transmitted by any means without the written permission of the author.

Published by AuthorHouse   05/31/2018

ISBN: 978-1-5462-9082-7 (sc)
ISBN: 978-1-5462-9081-0 (e)

Print information available on the last page.

Any people depicted in stock imagery provided by Getty Images are models, and such images are being used for illustrative purposes only.
Certain stock imagery © Getty Images.

This book is printed on acid-free paper.

Because of the dynamic nature of the Internet, any web addresses or links contained in this book may have changed since publication and may no longer be valid. The views expressed in this work are solely those of the author and do not necessarily reflect the views of the publisher, and the publisher hereby disclaims any responsibility for them.

CONTENTS

Preface v
Acknowledgements vi
PART 1 1
Tests for Forms 1 and 2 1
    Test 1 2
    Test 2 5
    Test 3 8
    Test 4 12
    Test 5 16
    Test 6 19
    Test 7 22
    Test 8 26
    Test 9 30
    Test 10 34
    Solutions for Tests 1 – 3 38
        Test 1 38
        Test 2 41
        Test 3 45
    Answers from Tests 4 – 10 50
        Test 4 50
        Test 5 50
        Test 6 51
        Test 7 51
        Test 8 52
        Test 9 53
        Test 10 53
PART 2 56
Tests for Forms 3 And 4 56
    Test 1 57
    Test 2 61
    Test 3 65
    Test 4 69
    Test 5 74
    Test 6 79
    Test 7 84
    Test 8 89
    Test 9 94
    Test 10 99
    Solutions for Tests 1 And 2 105
        Test 1 105
        Test 2 111
    Answers from Test 3 – 10 118
        Test 3 118
        Test 4 119
        Test 5 121
        Test 6 122

- Test 7 124
- Test 8 125
- Test 9 125
- Test 10 127

PART 3 130
Examination Papers 130
- Examination Paper 1 131
- Examination Paper 2 136
- Examination Paper 3 140
- Examination Paper 4 144
- Examination Paper 5 149
- Examination Paper 6 154
- Examination Paper 7 160
- Examination Paper 8 165
- Examination Paper 9 170
- Examination Paper 10 175
- Examination Paper 11 180
- Examination Paper 12 185
- Examination Paper 13 190
- Examination Paper 14 195
- Examination Paper 15 200
- Answers for Examination Papers 1-15 205
  - Examination Paper 1 205
  - Examination Paper 2 206
  - Examination Paper 3 207
  - Examination Paper 4 209
  - Examination Paper 5 209
  - Examination Paper 6 210
  - Examination Paper 7 211
  - Examination Paper 8 212
  - Examination Paper 9 213
  - Examination Paper 10 214
  - Examination Paper 11 216
  - Examination Paper 12 216
  - Examination Paper 13 217
  - Examination Paper 14 217
  - Examination Paper 15 218

Preface

Test Your Mind for Completing Ordinary Level Mathematics has been written to meet the needs of ordinary level students. It contains different kinds of questions from the "**Ordinary Level Syllabus**" prepared by the Tanzania Institute of Education in 2005. The book is divided into the following three parts:

- **part 1 (Tests for forms 1 and 2)**, which contains ten tests with twenty-five questions each for preparing forms 1 and 2 students
- **part 2 (Tests for forms 3 and 4)**, which contains ten tests with twenty-five questions each for preparing forms 3 and 4 students.
- **part 3 (Examination papers)**, which contains fifteen examinations with sixteen questions each for preparing the students for their final national examinations.

Therefore, the book has the total of **740** questions. Five tests (three tests from part 1 and two tests from part 2) have been done just as examples to prepare the students to do the remaining tests and examinations. Answers have been given at the end of each part.

It is hoped that students and teachers will find the book a nice aid for a better understanding of mathematics.

Acknowledgements

As a book writer, I have taken so much effort to come to this stage. However, it would also not have been possible to make the work a success without the kind support and help of many individuals. It is difficult to mention them all, but there are some who were great contributors and played an important role in the preparation of this book. With respect to their contributions, I must mention a few of them.

First of all, I should thank **God**, the Most Merciful, for His permission, which allowed me to complete this book.

My sincere thanks and appreciation go to my parents, Ramadhan Kombo Feruz and Fatma Omar Juma, for educating me and showing me a way to become a better person. I thank them for their great support. Once again, after many years, I am able to publish the book.

It would be a paucity of gratitude without mentioning my aunt, Zakiya Omar Juma. She supported me in the struggle of finding a sponsor.

My great thanks should go to my teachers, Mr. Abdalla Suleiman who taught me basic mathematics at Fidel Castro High School, helped me to edit the book, and Mr. Hadi, who taught me advanced mathematics at SOS Herman Gmeiner Secondary School Zanzibar, advised me regarding the name of the book.

I wish to thank all who assisted me, but whose names do not appear here. I appreciate their contributions.

PART 1

Tests for Forms 1 and 2

Test 1

Section A

1. Calculate the mean of the numbers 189, 207, 199, 285, 165, 158, and 183.

2. (a) Mention all prime numbers between 95 and 115.
   (b) Find the prime factors of 90.

3. If an operation * for any real numbers a and b is defined as $a*b = 2a^2 - b^3$, find
   (a) 2*5
   (b) (-2*3)*1

4. Solve the equation $\dfrac{7-x}{2} + \dfrac{2x+5}{3} = 6$.

5. (a) Simplify $\left(\dfrac{1}{2} \times 3\right) \div \dfrac{2}{5} + \dfrac{1}{4}$

   (b) $\dfrac{2}{3}$ of a cup contains 0.8 litres of milk. How many litres of milk can fill the cup?

6. Rationalize the denominator from $\dfrac{2\sqrt{3} - 2}{1 + \sqrt{3}}$.

7. Find the value of y in each of the following equivalent ratios:
   (a) $5:6 = y:2$
   (b) $\dfrac{1}{2} : \dfrac{1}{8} = y : \dfrac{3}{8}$

8. Draw the line of the equation $4x+y = 3$ on an x and y plane, and then state the x- and y- intercepts.

9. Calculate the area of figure.1.1.9 below.

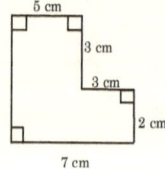

Figure 1.1.9.

10. (a) How many subsets does the set {a,b,c} have?
    (b) List all the subsets of the set in (a) above.

11. Calculate the value of the following without using any mathematical aids:
    (a) $\sqrt{12} \times \sqrt{27} \times \sqrt[3]{64}$
    (b) $\sqrt{\dfrac{7}{5}} \times \sqrt{\dfrac{343}{10}} \times \sqrt{2^{-1}}$

12. Divide the mass (29 g, and 20 cg) by 5.

13. Write $0.4\dot{5}\dot{7}$ in the form of $\dfrac{a}{b}$ where a and b are integers and $b \neq 0$.

14. What are the next two terms in each of the sequences below?
    (a) 6, 9, 14, 21, __ , __ .
    (b) 2520, 360, 60, 12, __ , __ .

15. Simplify the following without using tables:
    (a) $\log_9 3$
    (b) $\dfrac{1}{2}\log 625 + \dfrac{1}{3}\log 64$

16. (a) Find the value of x if $\left(\dfrac{1}{3}\right)^x \times \left(\dfrac{1}{27}\right)^{3x} = 9$
    (b) Factorize $8a^2 - 18b^2$ completely.

17. Solve the equation
    $$\tan 45° + x(\cos 60°) = 3,$$
    without using mathematical tables.

18. (a) Find the simple interest charged on \$125,000 when invested in a bank at 2.5% per annum for three years.
    (b) Make P the subject of the formula if $I = \dfrac{PRT}{100}$.

19. Find the image of point P(3,2) when reflected along the line $y = -2$.

20. The LCM of numbers 15, 30 and x is 90. Find the possible value(s) of x.

Section B

21. (a) Solve the quadratic equation $4x^2 + 8x = 14$ by completing the square.
    (b) Use mathematical tables to find the value of $\log_3 8$.

22. Find the area of the shaded region on the following semicircle if O is the centre of the circle. (Use $\pi = \dfrac{22}{7}$)

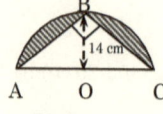

Figure 1.1.22.

23. Find the value of x from figure 1.1.23 below.

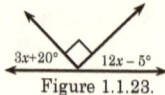

Figure 1.1.23.

24. Calculate the circumference of the circle (figure 1.1.24) below such that chords AB and CD are parallel; they are 8 cm apart; $\overline{CD} = 30$ cm; and O is the centre of the circle.

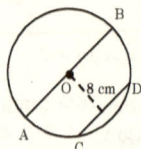

Figure 1.1.24.

25. The following data shows the scores of twenty-four students in a certain exam.

    20 32 18 23 29 18 19 18 20 18 23 20

    19 18 32 19 32 19 28 23 26 20 25 26

Use the grouping data method with class intervals 17–20, 21–24, etc. to represent the information above in a frequency distribution table indicating real limits, frequencies and class marks.

Test 2

Section A

1. (a) Write all the prime numbers between 120 and 140.
   (b) Find the LCM and GCF of numbers 24 and 30.

2. Multiply the time (5 hours, 20 minutes, and 50 seconds) by 20.

3. The operation $x*y$ is defined as $(x+y)^2$ for all real numbers x and y. Find $3\sqrt{3} * 5\sqrt{3}$.

4. Solve the equation $\dfrac{20}{2x-2} - \dfrac{25}{14-3x} = 0$

5. Simplify the following expressions:
   (a) $\dfrac{9x^2 y^{-5}}{2y^{-2} x^7}$
   (b) $\dfrac{2ab^2 \times 4a^5 b^2}{2a^2 b \div 3(ab)^3}$

   Write the answers with positive exponents.

6. Rationalize the denominator $\dfrac{a\sqrt{b} + b\sqrt{a}}{b\sqrt{a} - a\sqrt{b}}$

7. The degree measures of two complementary angles are in the ratio 7:11. Find the degree measure by each angle.

8. Determine the gradient, x-intercept and y-intercept from the equation $5x+4y = 8$.

9. What is the ratio x:y from figure 1.2.9?

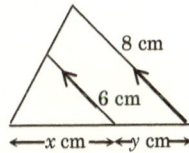

Figure 1.2.9.

10. If the universal set ξ = {a,b,c,d,e,f,g,h} and its subsets A = {a,b,d,f} and B = {b,c,d,g}, find the subsets
    (a) A∪B
    (b) A∩B
    (c) (A'∩B)'

11. Given that, $4x^2+y = \dfrac{16x^4 - y^2}{2x + \sqrt{y}}$, express y in terms of x.

12. Find the value of x and y from the figure 1.2.12 if $\overline{AC}$ and $\overline{EH}$ are straight lines.

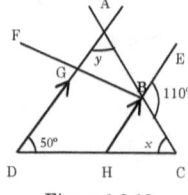

Figure 1.2.12.

13. (a) If $2a \times 10^2 + 3(a \times 10)^2 = 1.6 \times 10^3$, find the value of a.
    (b) Without using tables evaluate $22.5^2 - 12.5 \times 14.5$

14. Use the figure 1.2.14 to show that $x = y - z$.

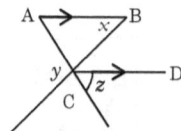

Figure 1.2.14.

15. The sum of three consecutive odd numbers is equal to the difference between 84 and $\dfrac{1}{9}$ of the middle number. Find the numbers.

16. Write the next two numbers in each of the following sequences:
    (a) 4, 9, 16, __, __.
    (b) 36, 81, 196, 441, __, __.

17. Maryam made the chips weigh 6.5 kg, and then she put them into packets. In each packet, she put 13 g of the chips. How many packets did Maryam get?

18. (a) How many minutes are there in $\dfrac{1}{8}$ of the month of June?
    (b) How many seconds in a week?

19. (a) Factorize completely the expression $-4y^2+16$
    (b) Use the factors of $-4y^2+16$ to solve equation $-4y^2+16 = 0$

20. Find the number of angles of a regular polygon whose interior angle is
    (a) 140°
    (b) 156°

Section B

21. (a) Given that $A = \dfrac{h}{2}(a+b) + \dfrac{\pi r^2}{2}$, make r the subject of the formula.
    (b) From the formula in (a) above, calculate r if $A = 100$, $a = 8$, $b = 10$ and $h = 6$.

22. Find the perimeter and area of figure 1.2.22 below if BD = 25 cm, BC = 15 cm, and the segment AB is the semicircle.

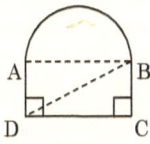

Figure 1.2.22.

23. The area of figure 1.2.23 below is 40.5 cm². Calculate its perimeter.

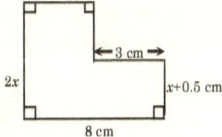

Figure 1.2.23.

24. Use prime factors to find the cube roots of the following numbers:
    (a) 16003.008
    (b) 3061.257408

25. Use mathematical tables to calculate the values of the following:
    (a) $\dfrac{25.4 \times \sqrt{15.6}}{32.12}$
    (b) $\dfrac{\sqrt[3]{82.17}}{\sqrt{15.32}}$

Test 3

Section A

1. Without using mathematical tables or a calculator, find the value of
   (a) $\sqrt{1.44}$
   (b) $\sqrt{14} \times \sqrt{\dfrac{49}{3.5}}$

2. The LCM of numbers 8, 12, 16, and w is 240. If w is less than 40, find the possible value(s) of w.

3. (a) Find the value of n in the line $nx - y = 0$ if it passes through point $(2,4)$.
   (b) The x- and y- intercepts of a line are 2 and 3 respectively. Find the equation in the form of $ax+by+c = 0$ where a, b and c are integers.

4. Simplify the following expressions:
   (a) $\dfrac{5a \times 4b - 2ab^2 + 8a^2b \div 2a}{2ab}$
   (b) $\dfrac{x^2 - y^2}{(x+y)^2}$

5. Solve the following simultaneous equations:
   $$2y-3x = 4x-\dfrac{3}{2}y = 3$$

6. Without using any mathematical aid, evaluate
   $$\dfrac{2\sqrt{3} - 3\sqrt{2}}{2\sqrt{3} + 3\sqrt{2}}$$
   given that $\sqrt{6} = 2.45$

7. The degree measures of interior angles of a trapezium are in the ratio 1:2:4:5. Find the degree measure of each angle.

8. (a) Divide the time (9 hours, 1 minute) by 4
   (b) How many seconds in $\dfrac{2}{3}$ of a day?

9. (a) If $x{:}y = 2{:}5$ and $y{:}z = 3{:}8$, find the ratio $x{:}y{:}z$.
   (b) Solve the inequality $|x{-}2| \geq 8$, and then represent the answer on the number line.

10. By using the following Venn diagram, find the value of x if $n(\xi) = 62$ where ξ is a universal set.

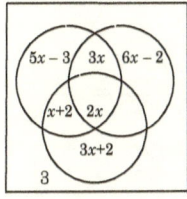

Fig.1.3.10.

11. If $\dfrac{1}{a} - \dfrac{1}{b} = \dfrac{1}{2}$ and $b - a = c$, find a in terms of c only.

12. The ratio of two positive consecutive numbers differing by 5 is 4:5. Find the sum of the numbers.

13. Factorize completely the following expressions:
    (a) $2a^2 - ab - b^2$
    (b) $9 - (x - 1)^2$

14. A number of two digits has the property in which the digit whose value is ones is 6 greater than that whose value is tens. If the sum of those digits is 8, find the numbers.

15. When the area of a small square whose one of the side is 4 cm is removed from the area of a large square, the remaining part is $9\,cm^2$. Find the lengths of the sides of the large square.

16. Make r_2 the subject of the formula $\dfrac{1}{r} = \dfrac{1}{r_1} + \dfrac{1}{r_2} + \dfrac{1}{r_3}$.

17. 10 people can do a job in 4 days. How many days can 8 people do the same job?

18. (a) Show that $\log_b a = \dfrac{1}{\log_a b}$

    (b) Find x from the logarithmic equation
    $$\log_b x = \frac{1}{3}\log_b 8 + \frac{1}{5}\log_b 32$$

19. Find the image of a point $A(3, 4)$ when reflected along
    (a) the x-axis
    (b) the line $x = -3$.

Tests for forms 1 and 2

20. On the figure 1.3.20, prove that $c = 180° - a - b$.

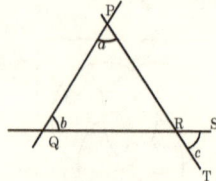

Figure 1.3.20.

Section B

21. Calculate the area and perimeter of figure 1.3.21 below given that FE = 4 cm, ED = 20 cm, AB = 5 cm, BC = 15 cm, and CD = 20 cm.

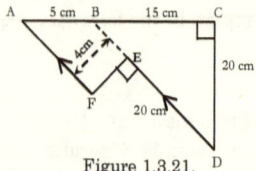

Figure 1.3.21.

22. (a) Find the value of y from figure 1.3.22 below if O is the centre of the circle.

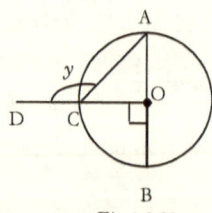

Fig.1.3.22.

(b) The exterior angle of a regular polygon is 60°. How many sides does that polygon have? Name it.

23. Find the volume of figure 1.3.23 if the radius of the base is 7 cm. Convert the answer in litres correct to 3 significant figures.

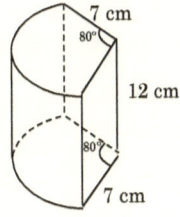

Figure 1.3.23.

24. Find the value of x and y from the figure 1.3.24 below.

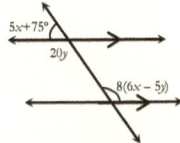

Figure 1.3.24.

25. (a) Find the equation of a line whose gradient is -4 and crossing the x-axis at 3. Hence state the y-intercept.
    (b) The slope of the line $ay+(5a+2)x - 4 = 0$ is 3. Find the value of a, and hence determine the y-intercept of the line.

Tests for forms 1 and 2

Test 4

Section A

1. The GCF of three numbers is 12. If two of the numbers can be expressed as $2^3 \times 3^2$ and $2^4 \times 3$; and the third number is a multiple of 60 and less than 240; find the possible value(s) of the third number, and hence find the LCM of those three numbers.

2. Solve for y from the equation
$$\log y = \log 216 + \frac{1}{3}\log 125 - 4\log 3 - \log \frac{16}{3}$$

3. A shopkeeper gets 115% by selling the rice at shs.2875 per kilogram. At what price would he need to sell 3.5kg of rice to make a profit of 30%?

4. Divide the time (89 hours, 11 minutes, and 8 seconds) by 44.

5. Calculate the rate of interest charged when $65,000 earns $2,600 in two years and eight months.

6. (a) If $\left(\frac{1}{4}\right)^{2n+2} = \left(\frac{1}{8}\right)^{3n}$, find the value of n.
   (b) Evaluate $\dfrac{\sqrt[3]{91.125}}{\sqrt{2.25}}$

7. Asha, Ali and Juma shared their oranges in the ratio 2:3:4 respectively. If Ali got nine oranges,
   (a) how many oranges did they share?
   (b) find the share of Asha and Juma.

8. Set µ is a universal set such that µ = {Odd numbers between 5 and 45} and its subsets A = {Prime numbers greater than 9} and B = {Multiples of 3}. Find
   (a) A∩B
   (b) (A∪B′)′
   (c) (A∪B)∩A′

9. The area of a circle was 616 cm². After enlarged, its radius has now increased by 25%. What is the area and circumference of the new circle? (Use $\pi = \frac{22}{7}$).

10. Given that $\dfrac{2a+b}{3a-2b} = 5$, find the ratio $a:b$.

11. The length of two parallel sides of a trapezium are in the ratio 3:7 and its height is 8 cm long. If the area of that trapezium is 160 cm², find the lengths of its parallel sides.

12. Given that $\log a = 2.5$ and $\log b = 4.5$. Find

   (a) $\log \dfrac{a\sqrt{b}}{b\sqrt{a}}$

   (b) $\log \dfrac{\sqrt[3]{ab^2}}{\sqrt{ab}}$

13. The product of two-third and the sum of a certain number and 5 is the difference between 5 and the number. Find the number.

14. (a) Factorize $\cos^2 A - \sin^2 B$

   (b) Use the result obtained in (a) above to solve the equation

   $$\cos^2\theta - \sin^2 60° = 0$$

   if θ is an acute angle.

15. Find the area of triangle XOY such that X and Y are at x and y-axis respectively and O is the origin. Given that the equation of a line XY is $5y = 2x+50$

16. On the square hereunder (figure 1.4.16), prove that $x = y\sqrt{\dfrac{1}{2}}$

   Hence show that the area of this square is given by $A=\dfrac{1}{2}y^2$

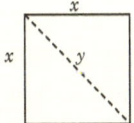

Figure 1.4.16.

17. Simplify the following expressions:

   (a) $\sqrt{4x+8y} + \sqrt{16x+32y}$

   (b) $\sqrt[3]{a^8} \div \sqrt[3]{8a^2}$

18. Solve the inequality $|x+0.5| \geq 2$ and show the answer on the number line.

19. (a) If the point $(2,n)$ is on the line $y+x+6 = 0$, find the value of n.
    (b) What is the condition that the simultaneous equations below have no solution?
    $$ax+by = p,\ cx+dx = q$$

20. Simplify the following expressions:
    (a) $2(2x - (3x - 4(2+x)) - 7x+2)$
    (b) $6x+2(3x - 3y)+(-3y(x - 5y+2))$

Section B

21. Calculate the area of figure 1.4.21 below. (Use $\pi = 3.14$)

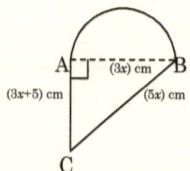

Figure 1.4.21.

(AB is a diameter of the circle)

22. From the following rectangle, find the value of x, and y, then calculate the perimeter of that rectangle.

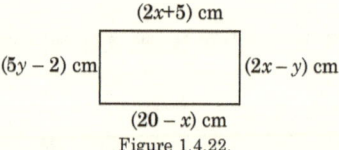

Figure 1.4.22.

23. Simplify the following expressions and write the answer in positive exponents.

(a) $\dfrac{12b^9 \times 3y^8}{2y^{-9} \times 3b^{-8}}$

(b) $\sqrt{\dfrac{(5x)^{13} \times (0.5y)^{-20}}{\left(\frac{0.2}{x}\right)^{-17} \times (16y^{-3})^3 \times y}}$

24. Find the area of the shaded region from the figure 1.4.24 below. (Use $\pi = \dfrac{22}{7}$)

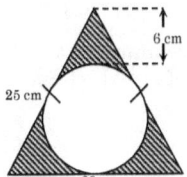

Figure 1.4.24.

25. (a) Make n the subject of the formula from $a = \dfrac{t-n}{\sqrt{t^2 - n^2}}$

(b) From the results in (a), find n if $t = 5$, $a^2+1 = 10$ and $a^4 - 1 = 40$.

Test 5

Section A

1. Multiply (4 hours, 24 minutes, and 25 seconds) by 115.

2. (a) Find the value of a, and b from the figure 1.5.2(a).

   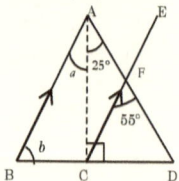

   Figure 1.5.2(a).

   (b) From parallelogram below, prove that $\triangle ABD \equiv \triangle CDB$, and hence show that $\angle BAD = \angle BCD$

   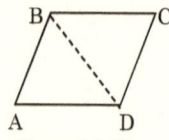

   Figure 1.5.2(b).

3. Given that log 2 = 0.3010, without using mathematical tables, find
   (a) log 5
   (b) log 0.025

4. Solve for x and y from the simultaneous equations
   $$2x+3y = 3 \text{ and } 8x^2 - 18y^2 = 54.$$

5. The ratio of Asha's and Ali's age is 2:3 respectively. Four years ago this ratio was 4:7. Find their present age.

6. By using a Venn diagram, or otherwise, show that
   (a) $(P' \cap Q')' = P \cup Q$
   (b) $P \cap Q = (P' \cup Q')'$

7. Solve the equation $3\dfrac{x}{4} = \dfrac{15-x}{2}$

8. Find the time period if the simple interest on £38,000 at 4.5% per annum is £8,550.

9. Given that x is an acute angle and $\tan x = \dfrac{3}{4}$, find the value of $\cos(2x)$ if,
   $$\cos(2x) = 1 - 2\sin 2(x).$$

10. Three bells are set to ring at intervals of 18 minutes, 24 minutes and 36 minutes. If they are all started together, how many times are they going to ring together on the same time in one day?

11. Six pipes can fill a tank of 500 litres for 2 hours. How many pipes of the same size can fill the tank for 6 hours?

12. A translation T takes point P(3,2) to P'(7,-3). Find
    (a) the translation T.
    (b) where will T take the point Q(5,-4).

13. Without using tables, write down the value of

$$\frac{5\sin 17°}{\cos 73°} + \frac{4\cos 35°}{\sin 55°}$$

14. Factorize completely the following expressions:
    (a) $a^2+b^2+3a+2ab+3b$
    (b) $a^2+b^2-c^2-2ab$

15. The magnitude of the area of the circle is the same as the magnitude of its radius (ignoring the units). Find the circumference of that circle. (The units are in centimetres).

16. A certain amount of money was divided to three people in the ratio 1:2:3. The difference between the largest and the smallest share is $10,000. How much did each one get?

17. Find the dimensions of a rectangle whose perimeter and area are 30 cm and 56 cm² respectively.

18. Johnny spent $\frac{3}{5}$ of his money on rent, $\frac{7}{16}$ of the remainder on clothes, $\frac{4}{15}$ of the remainder on food. If £1650 left, how much money did he have before the spending?

19. Write the next two terms in each of the sequences below:
    (a) 81, 49, 16, 0, __, __.
    (b) $\frac{1}{9}, \frac{1}{27}, \frac{1}{81}$, __, __ .

20. (a) Show that $(a+b)^2 = (a-b)^2 + 4ab$
    (b) Given that $x = L + \dfrac{t_1 c}{t_1 + t_2}$, make t_1 the subject of the formula.

Section B

21. Find the area and perimeter of the following figure.

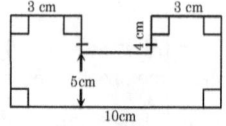

Figure 1.5.21.

22. (a) The number of animals on a farm are recorded as follows:

Cows	Goats	Sheep	Pigs
15	30	24	21

Construct a pie chart to represent the data above.

(b) From the table in (a) above, find the percentage of number of goats in the farm.

23. (a) Find the value of x if the area of the triangle below is 22 cm².

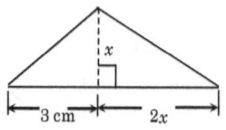

Figure 1.5.23.

(b) Jane is on the top of a building of height 20 m, looking down at a boat which is 100 m out to the sea from the base of the building. What is the angle of depression of the boat from the top of the building?

24. (a) Draw the graph of $y = 7x - 3 - 2x^2$, hence find the value of x from the equation $7x - 3 - 2x^2 = 0$.

(b) Use the graph you have drawn in (a) above to solve the equation $2x^2 = 6x - 4$.

25. Find the shaded area of figure 1.5.25 if the diameter, AC = 14 cm, and AB = BC. Take $\pi = \dfrac{22}{7}$.

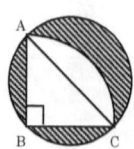

Figure 1.5.25.

Test 6

Section A

1. Divide the mass (493 kg, 948 g, and 620 mg) by 44.

2. If the universal set µ = {1,2,3,4,...10}, and its subsets A = {2,6,7,9} and B = {1,2,3,...6}, find
   (a) A'
   (b) A'∪B'
   (c) (A∩B')'∪(A'∪B)'. What does this set represent?

3. Solve the following simultaneous equations:
$$\frac{a}{b} + \frac{b}{2} = 2(a - \frac{3b}{2}) = 4$$

4. Find the area of the square below if the length of the diagonal is 10 cm.

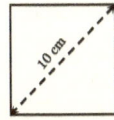

Figure 1.6.4.

5. The length and width of a certain rectangle are in the ratio 2:3. If its perimeter is 50 cm, find its area.

6. Solve for x from the logarithmic equation below:
$$\log_3 x + 2\log_x 3 = 3$$

7. The age of the father is now two times as the age as his son. Four years ago, a father was ten years older than the sum of the present age of his son and six years. Find the present ages of the father and his son.

8. Jane goes to supermarket for doing her business every day and she stays there for 12 hours a day. In September 2004, she was sick, and so she didn't go to the supermarket in the whole month. How long did she stay in the supermarket in that year?
   (Write the answer in seconds).

9. Factorize completely the following expressions:
   (a) $6xy+5x^2+y^2$
   (b) $a^4 - b^4$

10. Simplify the following:

$$\frac{\left(\frac{1}{x-3}+\frac{3}{x-1}\right)}{\left(\frac{2}{x^2-4x+3}\right)}$$

11. The perimeter of a quarter of a circle is equal to the sum of twice the radius and π. Find the circumference of the whole circle. (The units are in centimetres).

12. Given that $\dfrac{ab^2}{c}+c+\sqrt{ab}=0$, make c the subject of the formula.

13. If $a*b = 2a - b^2$, find
    (a) $(3*2)*5$
    (b) x if $[(1*x)*3] = -23$

14. Convert (a) 2.5 g/cm³ into kg/m³.
             (b) 2.5 litres into mm³.

15. Given that three points $(-2,4)$, $(2,2)$ and $(x,-1)$ are collinear. Find the value of x.

16. Find the possible values of x from the equation

    $|-4x+0.5| = 8.5$

17. The factors of a quadratic equation $3x^2 - ax+2 = 0$ are $(bx - 1)$ and $(x - 2)$. Find the possible values of a and b.

18. Given that $2^a = 1.5$, without using any mathematical aid, evaluate $\dfrac{4^{2a-2}}{8^{a-2}}$

19. Find the value of a, b and c from figure 1.6.19 below if a+c = 15.

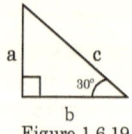

Figure 1.6.19.

20. Two doors are similar with a scale factor of 3. If the area of the smaller door is 4 m², find the area of the larger door.

Section B

21. Find the perimeter and area of figure 1.6.21 shown below.

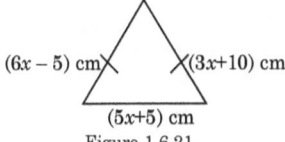
Figure 1.6.21.

22. Evaluate the following by using mathematical tables. Give the answer in standard form to 3 significant figures.

$$\frac{\sqrt{25.35} \times \sqrt[3]{15.25}}{\sqrt[4]{43.85} \times \sqrt[3]{34.45}}$$

23. On the figure 1.6.23 below, find the values of x and y.

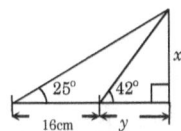
Figure 1.6.23.

24. (a) Solve for x and y if $\dfrac{3^{3y} \times 5^{2x}}{15^{-x} \times 3^{6y}} = 45$

(b) Rationalize the denominator from $\dfrac{2\sqrt{3}+2}{\sqrt{3}-4}$

25. (a) The point (5,3) is reflected to the point (-2,3). What is the equation of the line of reflection?

(b) Plot the image of triangle ABC with vertices A(1,2), B(2,1) and C(2,2) after rotation of 90° anticlockwise direction about the origin.

(c) The scale of a certain map is 1:25,000. If the distance from point A to B on the map is 4.5 cm, calculate the distance from A to B on the ground in kilometers.

Test 7

Section A

1. (a) Add: (5 km, and 9 hm) by (7 km, and 70 dm). Write the answer in dm.
   (b) When the tank is filled with water, it weighs 30 kg, and 50 cg. But when the tank is empty, it weighs 5 kg, and 59 cg. What is the mass of water that the tank can take?

2. Solve the following simultaneous equations:

$$3x+3y = y+3z = 8x - \frac{1}{4}z = 15$$

3. Find the value of x on the figure 1.7.3 below.

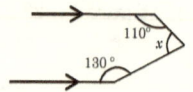

Figure 1.7.3.

4. Solve the equation $3\frac{x}{3} + 5 = \frac{4x+18}{2}$

5. Simplify

$$\frac{15ab^2c^4 + 10a^4b^2c - 5(abc)^2}{ac - 3c^3 - 2a^3}$$

6. The area of a rectangle was 20 cm² and its length was 5 cm. If now each of its sides has increased by 50%, what is its area now?

7. Juma bought x loaves of breads for shs.50 each. Those loaves would be two more, than if he would buy them for shs.70 each. Find the value of x.

8. Jacky has fifteen balls. Each ball has white, blue or both colours. If eight balls have white colours, four balls have both white and blue colours and x balls have blue colours. Find the value of x.

9. Calculate the amount of money that will be accumulated when £65,700 is invested in a bank at 3.5% per annum at simple interest, after five years and six months.

10. (a) Factorize completely $49(a+b)^2 - 25(b-2a)^2$
    (b) Evaluate $(2589.25)^2 - (2589.25 \times 2489.25)$

Tests for forms 1 and 2

11. State whether the following sets are finite or infinite:
    (a) A = {All digits}
    (b) B = {$x: 0 < x < 2$}
    (c) C = {All negative integers less than -100}
    (d) D = {All negative integers greater than -100}

12. The LCM of three numbers is 720 and their GCF is 12. If two of the numbers are 36 and 48, find the third possible number(s).

13. Find the value of θ in each of the following equations without using tables if θ is an acute angle.
    (a) $\cos \theta = \sin 38°$
    (b) $\tan \theta = \dfrac{\cos 15°}{\sin 15°}$

14. Given that $\log 2 = 0.3010$, evaluate $\log \dfrac{1}{\sqrt{8}}$ without using mathematical tables.

15. (a) Convert $0.08\dot{3}$ into fraction.
    (b) Round of 0.0748 to the nearest thousandth

16. The area of a sector of a circle at 60°, having the square of its radius 73.5 cm², is equal to the area of 25% of the circle whose radius is r. Find r.

17. Find $E\hat{C}F$ from the figure 1.7.17 below, if $B\hat{A}F = B\hat{F}A$ and $\overline{BC} = \overline{CF}$.

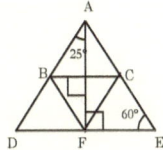

Figure 1.7.17.

18. If $x*y = 2(x^2 - 4y^2)$, find the value of
    (a) $(1*2)*(1*-2)$
    (b) n if $n*2 = 0$

19. Find the area of the rhombus below.

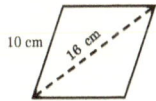

Figure 1.7.19.

20. Simplify the following by using the laws of exponents if $(a^2+1) = 5$

$$\frac{a^{12} - 2^3 + a^{16} + a^{14} + 15}{3a^{12} + 1}$$

Section B

21. (a) If $\tan A = 0.75$, find the value of

$$\frac{\sin A + 2\cos A}{\cos A - \sin A}$$

(b) Show that $\tan\theta \tan(90° - \theta) = 1$

22. The length of a rectangle is $(2x+5)$ cm and its width is $(4x-4)$ cm. If its perimeter is 0.62 m, find the area of the rectangle.

23. Find the area and perimeter of the following triangle.

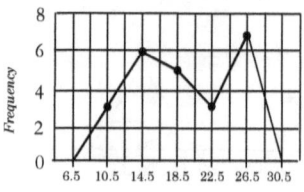

Figure 1.7.23.

24. The following figure is a frequency polygon which represents the ages of pupils in years in a certain class. Study the frequency polygon and answer the questions.

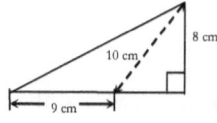

class mark

Figure 1.7.24.

(a) How many pupils have the ages between class marks 18.5 and 26.5 inclusive?

(b) What is the class interval of class mark 14.5?

(c) How many pupils are in the class?

(d) Construct the frequency distribution table and represent all the information from the frequency polygon above.

25. An iron sphere shown on figure 1.7.6 below is to be melted and used to make a wire. If the radius of the wire is to be 0.14 cm, find the total length of the wire that can be made using this material.

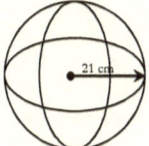

Figure 1.7.25.

The radius of the sphere is 21 cm.

Tests for forms 1 and 2

Test 8

Section A

1. (a) Multiply the mass (5 kg, 125 g, and 149 mg) by 58.
   (b) How many seconds are there in $\frac{3}{8}$ of the year 2005?

2. Simplify the following expression:
$$\frac{2}{5a^2 - 20a - 60} \div \frac{2a+6}{a^3 - 3a^2 - 18a}.$$

3. A farmer sold half a number of the oranges and then he sold a quarter of the number of the oranges left. He also sold one-third a number of the mangoes and then added a quarter of the number of mangoes left. What is the total number of oranges and mangoes altogether does the farmer have by now?

   (Take x as the original number of oranges and y as the original number of mangoes)

4. In a class of thirty students, sixteen of them play football, thirteen play basketball and five play both football and basketball. How many students play neither football nor basketball?

5. (a) Solve for y, given that $\log_3 y - \log_3 (y-8) = 2$
   (b) Without using mathematical tables, evaluate

   $(0.2\log 1024) - (0.5\log 4) + \frac{1}{3}\log 125$

6. The magnitude of the circumference of a circle is equal to the magnitude of its area (ignoring the units). Find its radius if the units are in metres.

7. The medium-sized angle of a triangle exceeds the smallest angle by $40°$ and the largest one is twice the medium-sized angle. Find the size of each angle.

8. Find the product of (a) $\bar{7}.8975 \times 12$ (b) $\bar{11}.3457 \times 15$

9. (a) Show that $(a+b)^2 - (a-b)^2 = 4ab$.
   (b) Find the numerical value of $3a^{8x} - 2a^{6x}$ if $a^{2x} = 2$.

10. The points (5,-2) and (3,2) lie on the line $\frac{x}{a} + \frac{y}{b} = 1$ when a and b are any real numbers. Find the value of a and b.

Tests for forms 1 and 2

11. (a) Factorize completely the expression

    $3a^2c - 5a^2d - 3b^2c + 5db^2$.

    (b) Without using any mathematical aid, evaluate

    324×(-324)+376×376

12. The scale of a map is 1:250,000. Calculate
    (a) the actual distance in kilometres between two points A and B which are 8cm apart on the map.
    (b) the distance on the map which represents 4.2 km on the ground.

13. The operation * is defined on a set of real numbers as:

    $$x * y = \begin{cases} 4 - x \text{ if } x \geq y \\ y + 2 \text{ if } x < y \end{cases}$$

    Find the value of
    (a) (2*3)*5
    (b) a if $2a*3 = 2$ given that $a > 1.5$

14. Solve the following equation:

    $$2y + \sqrt{2y - 2} = 2$$

15. The area of a rhombus whose diagonals are in the ratio 3:4 is 54 m². Calculate
    (a) the lengths of the diagonals.
    (b) the perimeter of the rhombus.

16. (a) If $a^y = b^x$, find the ratio $x:y$.
    (b) Write the ratio 25.8:606.3 in the form of 1:n where n is any real number.

17. Use the factors of the expression $x^2 - 4y^2$ to solve the pair of simultaneous equations

    $$\begin{cases} x^2 - 4y^2 = 20 \\ x - 2y = 2 \end{cases}$$

18. Find the value of u, v and w, given that,

    $$\frac{u}{4} = \frac{v}{3} = \frac{w}{6} \text{ and } 3u + 3v - 4w = 24.$$

19. (a) If $\left(x+\dfrac{1}{x}\right)^2 = 5$, find the value of $x^2 + x^{-2}$.

(b) From the figure 1.8.19, AB//CD, prove that $\triangle ABX \sim \triangle CDX$, hence calculate the length AX if AC = 10 cm, AB = 12 cm and CD = 8 cm.

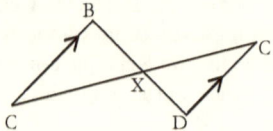

Figure 1.8.19.

20. The triangle ABC whose vertices are at A(2,3), B(1,-2) and C(4,3) is first reflected along the y-axis, followed by the reflection along the line $y = x$. Find the coordinates of the final image of triangle ABC.

Section B

21. Given that $A = \{x \in \mathbb{Z}: 5 \leq x \leq 20\}$. By using elements listing method (roster method), find the following subsets of A.
    (a) $B = \{x: x$ is a multiple of 3$\}$
    (b) $C = \{x: x$ is a prime number$\}$
    (c) (i) B' (ii) $(B \cup C)'$ (iii) $(B' \cap C')$

22. (a) Multiply the length (2 km, 5 hm, and 8 dam) by 1.5
    (b) The mean of the heights of a pawpaw tree and an orange tree is 2 m, and 55 cm. The mean of the heights of that pawpaw tree and a banana tree is 2 m, 35 cm. Find the height of the banana tree if the height of the orange tree is 2 m, 45 cm.

23. By using mathematical tables, evaluate the following:

(a) $\left(\dfrac{53.8}{24.7} + \dfrac{48.75}{38.67}\right)^2$ (b) $\dfrac{18.75^2 - \sqrt{847.6}}{\sqrt[3]{148.36}}$

24. Calculate the area of figure below.

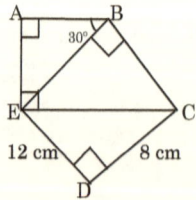

Figure 1.8.24.

25. The number of students in different classes of a certain school are shown on the table below.

class / sex	FI	FII	FIII	FIV	FV	FVI
Girls	28	30	24	20	12	14
Boys	26	25	16	18	15	18

Draw a bar graph to represent the information.

Test 9

Section A

1. (a) Solve for y from the following logarithmic equation:

   $$(\log_y 100)^2 + \log_y 10 = 5$$

   (b) Simplify the following expression:

   $$\log(a+b) + \frac{1}{2}\log(a-b) - \log 2 - \frac{1}{2}\log(a^2 - b^2)$$

2. On triangle ABC, $\angle BAC = 40°$; the side BC is produced to D such that $\overline{AC} = \overline{CD}$. Given that $\angle ABC = 2x + 20°$ and $\angle ADC = x$, calculate $\angle ACB$.

3. From the figure 1.9.3 below, prove that $a+b+c+d+e = 1.5(f+g+h+i+j)$

   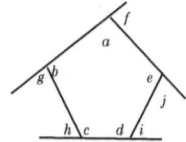

   Figure 1.9.3.

4. The LCM and GCF of three numbers are 25200 and 60 respectively. If two of the numbers are 2100 and 1800, find the possible value(s) of the third number.

5. (a) If the volume V, of a sphere is given by $V = \frac{4}{3}\pi r^3$ and the area is given by $A = 4\pi r^2$, find V in terms of A and π but not r.

   (b) If the area of a sphere, $A = 1256 cm^2$ and $\pi = 3.14$, find the volume of the sphere.

6. (a) Given that $n(X) = a$, $n(Y) = a+3$, $n(X \cap Y) = 4$ and $n(X \cup Y) = 15$ where X and Y are sets. Find the value of a.

   (b) If A and B are disjoint sets, and that $n(A') = 20$, $n(B') = 14$, and $n(\xi) = 30$ where ξ is a universal set, find $n(A \cup B)$.

7. Given that

   $$\frac{1}{x-1} + \frac{1}{x+1} = \frac{ax}{x^2 - b},$$

   for any real value of x (except for 1 and -1), find the possible value(s) of a and b.

8. Find two positive integers in the ratio 2:3 whose squares differs by 80.

9. Factorize completely the expression $(a-3)^3 + 3 - a$, and use the factors to solve the equation $(a-3)^3 + 3 - a = 0$.

10. Make r the subject of the formula in the equation $V = 2\pi r(r+h)$.

11. Given that $\dfrac{13.04 \times 10^2 \times 1.576 \times 10^6}{3.26 \times 10^4} = 6.304 \times 10^4$,

    without using tables find

    $$\dfrac{0.1304 \times 10^2 \times 15.76 \times 10^{-3}}{0.326 \times 10^4}.$$

12. If $[4^{(x-1)}].[3^{(y+3)}] = 3888$, find the possible values of x and y.

13. Two numbers a and b are chosen so that their sum is equal to thrice of their difference. Find the value of $\dfrac{ab}{a^2 + b^2}$.

14. Find the quotients of the following:
    (a) $\overline{1}0.5346 \div 9$
    (b) $\overline{1}3.3847 \div 15$

15. If the image of point P when reflected along the line $x = 2$ is P'(3,5). Write the coordinates of P.

16. The area of two circles are in the ratio 16:49. Calculate
    (a) the radius of the smaller circle if the diameter of the larger one is 28 cm.
    (b) the circumference of the smaller circle.

17. Express $x^2 - y \times z$ in terms of x and y only as the sum of product of two factors of the difference between two squares and y, if $y = z+1$. Use the result to evaluate $(1015)^2 - 915 \times 914$.

18. (a) The area of a circle is 15π greater than its circumference (ignoring the units). Find the radius of the circle.
    (b) A two digit number is chosen in such a way that it is 12 more than twice the sum of its digits. It is also seven-fourth of the number formed by interchanging the digits. Find that number.

19. Find the value of r if $r_1 = 3$ and $r_2 - r = 4$, from the equation $\dfrac{1}{r} = \dfrac{1}{r_1} + \dfrac{1}{r_2}$ where r is positive.

20. (a) An article was sold at $50,000. If the price included the Value Added Tax (VAT) at 25%, find the amount of VAT.
    (b) Mary sells an article for $550 and makes a percentage profit of 37.5%. What would be her percentage loss if she sells the article for $350?

Section B

21. The following pie chart represents the total of 2,160 different fruits collected by a farmer in a certain village.

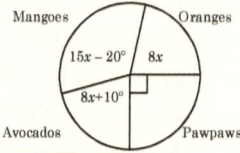

Figure 1.9.21.

(a) Find the value of x.
(b) Find the number of
   (i) mangoes   (ii) oranges   (iii) avocados
(c) What is the percentage number of pawpaws?

22. Find the volume of the figure 1.9.22 below

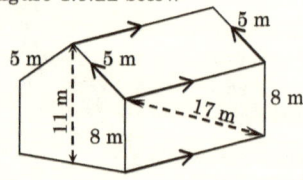

Figure 1.9.22.

23. On the figure 1.9.23 below, find
   (a) the area, and
   (b) the perimeter of the diagram.

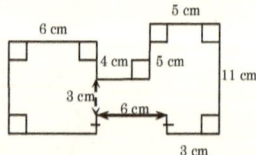

Figure 1.9.23.

24. (a) Solve graphically the simultaneous equations

$$\begin{cases} 2x + y = 11 \\ 3x - 2y = 6 \end{cases}$$

(b) Draw the graph of $y = x^2 - 3x$ for $\{-2 \leq x \leq 4\}$ and use the graph to solve the equation $x^2 - 3x - 4 = 0$.

25. (a) Round off 25.04, 69.56, 12.47, and 23.235 each to three significant figures, and hence use the results to evaluate

$$\frac{25.04 \times 69.56}{12.47 \times 23.235}$$

(b) Use mathematical tables to find the value of the following correct to 2 decimal places.

$$\frac{\sqrt{23.38} \times 48.67}{2\pi + (0.581)^2}$$

Test 10

Section A

1. If n(X) = 10, n(X∩Y) = 4, n(Y') = 9, and n(μ) = 15 where μ is a universal set, find
   (a) n(X')
   (b) n(X∪Y)
   (c) n(X'∩Y')

2. Evaluate the following:
   (a) $\log_{\frac{1}{2}}\left(\frac{1}{16}\right)$
   (b) $\log_{\frac{1}{3}} 81$.

3. If $a^2 = bc$ and $b = 5$, find c from $\log a + \log \dfrac{a^2}{c} = 2$

4. Books are sold in boxes of 15 books, pens in boxes of 21 pens and rulers in boxes of 24 rulers. What is the minimum number of boxes of each item should I buy in order to get the same number of three items?

5. Solve the following equation by using the general formula of quadratic equations.
   $$\frac{2x^2 - 1}{(0.5 - x)(3 + x)} = \frac{1}{4}$$

6. Simplify $\left(\dfrac{3^{(a+b)} \times 5^{2a} \times 27^{3b}}{3^{2a} \times 25^{(2b-a)}}\right) \div \left(\dfrac{25^{(a+b)}}{3^{(a-2)} \times 125^{b}}\right)$

7. Find the value of a and b for which,
   $4x^2 - ax + 4 = 2[b(x - 1)^2]$, for any value of x.

8. Students in a certain class are some study geography, some study history and some biology. Twenty-one students study geography, eight study all the three subjects, seventeen study history and biology and fifteen study history and geography. Each student study more than one subject. Draw a Venn diagram to represent the information and use it to find the number of students
   (a) who study biology and geography.
   (b) who study history or geography but not biology.
   (c) in the class.

9. Mrs. Salim bought pencils for shs. 10,000. She sold 70% of them at a profit of 30%, 50% of the remaining at a loss of 8% and the rest at the same price as she bought them. Find her total profit and its percentage.

10. Without using any mathematical aid, evaluate

$$\frac{1.8 \times 10^{-2}}{1.5 \times 10^{-3} + 0.45 \times 10^{-2}}$$

11. (a) Solve for x and y from the equation $3^{2x} \times 5^{3y} = 12$.

   (b) Simplify $\dfrac{64^{-\frac{2}{3}} + 27^{\frac{2}{3}}}{8^{-\frac{5}{3}}} + \dfrac{1}{100^{-\frac{1}{2}}}$

12. The volume of a frustum is given by

$$V = \frac{1}{3}\pi h(R^2 + Rr + r^2),$$

   make r the subject of the formula.

13. From the figure 1.10.13 below, ABCD and PQRC are the squares. Prove that $\overline{DR} = \overline{PB}$, then calculate $B\hat{R}D$ if $P\hat{B}C = 25°$.

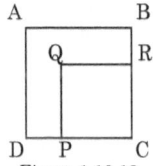

Figure 1.10.13.

14. In a certain village, there are 180 children. The ratio of the number of children, men and women in the village is 6:4:5 respectively. If the mean weight of the children and men are 22 kg and 86 kg respectively and the mean weight of the people in the whole village is 56.4 kg, find the mean weight of the women.

15. (a) Find xy if $x^2 + y^2 = 1000$ and $x - y = 30$.
    (b) Show that $2a^2 + 2b^2 = (a+b)^2 + (a-b)^2$.

16. (a) Solve the inequality $|2x+3| \leq 4$ and then represent the results for x on the number line.
    (b) The sum of a number and its reciprocal is 4.25. Determine the number.

17. (a) Write $0.08\dot{3}\dot{2}$ in the form of $\frac{a}{b}$ where a and b are any integers and $b \neq 0$.
    (b) Evaluate $\frac{25.8 \times 37.5}{54.3}$, hence correct the answer to the nearest tenth.

18. Two types of rice cost shs.800 and shs.1000 per kilogram each. In what ratio must the two be mixed so that the mixture costs shs.870 per kilogram?

19. What the principal will give the amount of £90,300 in three years at 2.5% per annum simple interest?

20. The total surface area of a cylindrical block is 297 cm². If the height of the block is 10 cm, find its radius (Use $\pi = \frac{22}{7}$)

Section B

21. The marks scored by forty students in a certain test were recorded as follows:

    38 14 37 26 56 36 60 28 30 44
    42 29 14 24 65 52 34 50 29 32
    27 50 08 35 38 55 36 42 37 48
    43 36 13 54 47 49 19 39 40 15

    (a) Prepare a frequency distribution table with class marks 5, 15, 25, etc.
    (b) Draw a cumulative frequency curve and use it to estimate the number of students scored at least 45.

22. (a) Plot a triangle with vertices at P(2,3), Q(0,3) and R(3,2). On the same axes, transform the triangle by the translation vector $\begin{pmatrix} 4 \\ -2 \end{pmatrix}$, then reflect the result in the line $x = -2$. Write the coordinates of the new vertices.
    (b) When 2.8 km on the land is measured on its map, it is found to be 2 cm. Calculate
    (i) the scale of the map.
    (ii) the actual distance in kilometres which is 1.2 cm on that map.

23. Calculate the perimeter of figure 1.10.23, if O is the centre of circle, AO = 20 cm, BO = 10 cm.

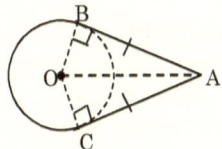

Figure 1.10.23.

24. (a) Solve the equation $2x^2 - 10 = 5x$ by completing the square.
    (b) The surface area of a cone is given by a formula $A = \pi r^2 + \pi r l$. Make l the subject, hence calculate l if $A = 2504.8$, $r = 12.8$ and $\pi = 3.14$, and write the answer to the nearest unit.

25. Given that $x = 0.8\dot{3}\dot{2}$ and $y = 1.04\dot{5}$. Find
    (a) in the same form (as recurring decimal),
      (i) $x+y$ (ii) $y - x$
    (b) $\dfrac{y^2 - x^2}{(y + x)^2}$ in the form of $\dfrac{a}{b}$ where a and b are integers and $b \neq 0$.

Tests for forms 1 and 2

Solutions for Tests 1 – 3

Test 1

1. Mean = $\dfrac{189+207+199+285+165+158+183}{7} = \dfrac{1386}{7} = 198$

2. (a) The odd numbers between 95 and 115 are:

    {97,99,101,103,105,107,109,111,113}

    -Multiples of 3 are {99,105,111}

    -Multiples of 5 are {105}

    -Multiples of 7 { }

    ∴ The prime numbers between 95 and 115 are

    97,101,103,107,109 and 113.

   (b) $90 = 2\times 45 = 2\times 3\times 15 = 2\times 3\times 3\times 5$

    ∴ The prime factors of 90 are 2, 3 and 5.

3. (a) $a*b = 2a^2 - b^3$

    $2*5 = 2(2)^2 - 5^3 = 8 - 125 = -117$

   (b) $-2*3 = 2(-2)^2 - 3^3 = 8 - 27 = -19$

    Now, $(-2*3)*1 = -19*1$

    $-19*1 = 2(-19)^2 - 1^3 = 722 - 1 = 721$

4. $\dfrac{7-x}{2} + \dfrac{2x+5}{3} = 6$

    $\dfrac{3(7-x)+2(2x+5)}{6} = 6 \Rightarrow 21 - 3x + 4x + 10 = 36$

    $x + 31 = 36 \Rightarrow x = 5$

5. (a) $\left(\dfrac{1}{2}\times 3\right) \div \dfrac{2}{5} + \dfrac{1}{4} = \left(\dfrac{3}{2} \div \dfrac{2}{5}\right) + \dfrac{1}{4} = \left(\dfrac{3}{2}\times\dfrac{5}{2}\right) + \dfrac{1}{4} = \dfrac{15}{4} + \dfrac{1}{4} = \dfrac{16}{4} = 4$

   (b) Let x litres of milk can fill the cup. Then; $\dfrac{2}{3}x = 0.8$

    $$x = \dfrac{0.8\times 3}{2} = 1.2 \text{ litres}$$

6. $\dfrac{2\sqrt{3}-2}{1+\sqrt{3}} = \dfrac{(2\sqrt{3}-2)(1-\sqrt{3})}{(1+\sqrt{3})(1-\sqrt{3})} = \dfrac{2\sqrt{3}-6-2+2\sqrt{3}}{1-3} = \dfrac{4\sqrt{3}-8}{-2} = 4 - 2\sqrt{3}$

7. (a) $5:6 = y:2 \Rightarrow \dfrac{5}{6} = \dfrac{y}{2} \Rightarrow y = \dfrac{5\times 2}{6} = \dfrac{5}{3}$

   (b) $\dfrac{1}{2} : \dfrac{1}{8} = y : \dfrac{3}{8} \Rightarrow \dfrac{1}{2}\times 8 = y\times\dfrac{8}{3} \Rightarrow y = \dfrac{4\times 3}{8} = \dfrac{3}{2}$

8.

x	-1	1	2
y	7	-1	-5

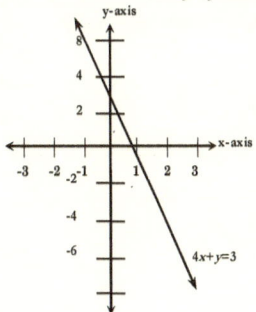

Then, x- and y- intercepts are 0.75 and 3 respectively.

9.

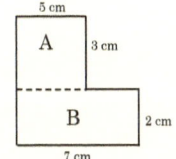

Total area = Area A+Area B

$= (5\times 3 \text{ cm}^2)+(7\times 2 \text{ cm}^2)$

$= 15 \text{ cm}^2+14 \text{ cm}^2$

$= 29 \text{ cm}^2$

10. (a) Number of subsets in a set is given by 2^n where n = 3

Then, $2^3 = 8$

∴The number of subsets is 8.

(b) $\{\},\{a\},\{b\},\{c\},\{a,b\},\{b,c\},\{a,c\}$ and $\{a,b,c\}$

11. (a) $\sqrt{12}\times\sqrt{27}\times\sqrt[3]{64} = \sqrt{4\times 3}\times\sqrt{9\times 3}\times\sqrt[3]{2^6} = 2\sqrt{3}\times 3\sqrt{3}\times 2^2 = 6\times 3\times 4 = 72$

(b) $\sqrt{\dfrac{7}{5}}\times\sqrt{\dfrac{343}{10}}\times\sqrt{2^{-1}} = \sqrt{\dfrac{7}{5}\times\dfrac{343}{10}\times\dfrac{1}{2}} = \sqrt{\dfrac{7\times 7^3}{5\times 5\times 2\times 2}} = \dfrac{7^2}{5\times 2}$

$= 49\div 10$
$= 4.9$

12.
```
   g      cg
5\ 29     20
   05     84
```
Note: 1 g = 100 cg

13. Let $x = 0.4\dot{5}\dot{7}$

$100x = 45.7\dot{5}\dot{7}$

$99x = 45.3$

$x = \dfrac{453}{990} = \dfrac{151}{330}$

14. (a) $6,9,14,21,x,y$

$\Rightarrow 6, 6+3, 9+5, 14+7$

$\therefore x = 21+9 = 30$

$y = 30+11 = 41$

(b) $360,60,12,x,y$

$\Rightarrow 360, 360\div 6, 60\div 5$

$\therefore x = 12\div 4 = 3$

$y = 3\div 3 = 1$

15. (a) $\log_9 3 = \log_9 \sqrt{9} = \frac{1}{2}\log_9 9 = \frac{1}{2}$

    (b) $\frac{1}{2}\log 625 + \frac{1}{3}\log 64 = \log 625^{\frac{1}{2}} + \log 64^{\frac{1}{3}} = \log 25 + \log 4$
    $$= \log 25 \times 4$$
    $$= \log 100 = 2$$

---

16. (a) $\left(\frac{1}{3}\right)^x \times \left(\frac{1}{27}\right)^{3x} = 9$

    $3^{-x} \cdot 3^{-9x} = 3^2$
    $3^{-x-9x} = 3^2$
    $-10x = 2$
    $x = -\frac{1}{5}$

    (b) $8a^2 - 18b^2 = 2(4a^2 - 9b^2)$
    $$= 2[(2a)^2 - (3b)^2]$$
    $$= 2(2a+3b)(2a-3b)$$

---

17. $\tan 45° + x\cos 60° = 3$
    $1 + x(0.5) = 3 \Rightarrow 0.5x = 3 - 1 = 2$
    $x = 4$

---

18. (a) $I = \frac{PRT}{100} = \frac{125{,}000 \times 2.5 \times 3}{100} = \$9{,}375$

    (b) $I = \frac{PRT}{100} \Rightarrow 100I = PRT \Rightarrow P = \frac{100I}{RT}$

19.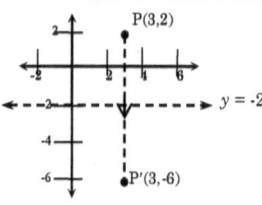

---

20. $90 = 2 \times 3 \times 3 \times 5$
    $15 = 3 \times 5$
    $30 = 2 \times 3 \times 5$
    $\therefore x = 3 \times 3 = 9$
    Or $= 2 \times 3 \times 3 = 18$
    Or $= 3 \times 3 \times 5 = 45$
    Or $= 2 \times 3 \times 3 \times 5 = 90$
    $\therefore$ The possible values of x are 9, 18, 45 and 90.

---

21. (a) $4x^2 + 8x = 14$
    $4(x^2 + 2x) = 14$
    $4(x^2 + 2x + 1^2) = 14 + 4$
    $4(x+1)^2 = 18$
    $x + 1 = \sqrt{\frac{18}{4}}$
    $x = \frac{-2 \pm 3\sqrt{2}}{2}$

    (b) $\log_3 8 = \frac{\log 8}{\log 3} = \frac{0.9031}{0.4771} = 1.893$

Number	Logarithm
0.9031	$\bar{1}.9557$
0.4771	$\bar{1}.6786$
1.893	0.2771

Tests for forms 1 and 2

22. Area of the semicircle = $\frac{1}{2}\pi r^2 = \frac{1}{2} \times \frac{22}{7} \times 14^2 = 308$ cm².

    Area of triangle ABC = $\frac{1}{2} \times 14 \times 28 = 196$ cm².

    ∴Shaded area = (308 − 196) cm² = 112 cm²

---

23. $3x+20°+90°+12x-5° = 180°$
    $15x+105° = 180°$
    $15x = 75°$
    $x = 5°$

24.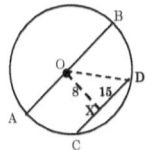

    Since CD = 30cm, then XD = 30÷2 = 15
    By Pythagoras theorem, $OD^2 = 8^2+15^2 \Rightarrow OD = 17$ cm
    Circumference = $2\pi r = 2 \times 3.14 \times 17$ cm = 106.76 cm.

25.

Class intervals	Tally mark	Frequency	Class mark	Real limits
17–20	//// //// ///	13	18.5	16.5-20.5
21–24	///	3	22.5	20.5-24.5
25–28	////	4	26.5	24.5-28.5
29–32	////	4	30.5	28.5-32.5

Test 2

1. (a) The odd numbers between 120 and 140 are:
    {121,123,125,127,129,131,133,135,137 and 139}
    -Multiples of 3 are {123,129,135}
    -Multiples of 5 are {125,135}
    -Multiples of 7 is {133}
    -Multiples of 11 is {121}
    ∴Prime numbers between 120 and 140 are 127,131,137 and 139.
   (b) 24=2×2×2×3, 30=2×3×5
    ∴LCM = 2×2×2×3×5 = 120
    and GCF = 2×3 = 6

---

2. h   min   s

    5   20   50   Note: 1 min = 60 s and 1 h = 60 min

    ×   20

    106 56 40

Tests for forms 1 and 2

3. $x * y = (x+y)^2$

    $3\sqrt{3} * 5\sqrt{3} = (3\sqrt{3} + 5\sqrt{3})^2 = (3\sqrt{3})^2 + 2 \times 3\sqrt{3} \times 5\sqrt{3} + (5\sqrt{3})^2$
    $= 27 + 90 + 75 = 192$

4. $\dfrac{20}{2x-2} = \dfrac{25}{14-3x} \Rightarrow 20(14-3x) = 25(2x-2)$

    $280 - 60x = 50x - 50$

    $280 + 50 = 50x + 60x \Rightarrow 110x = 330 \Rightarrow x = 3$

5. (a) $\dfrac{9x^2 y^{-5}}{2y^{-2} x^7} = 4.5(x^{2-7})(y^{-5+2}) = \dfrac{4.5}{x^5 y^3}$

    (b) $\dfrac{2ab^2 \times 4a^5 b^2}{2a^2 b + 3(ab)^3} = \dfrac{8a^6 b^4 \times 3a^3 b^3}{2a^2 b} = 12(a^{6+3-2})(b^{4+3-1}) = 12a^7 b^6$

6. $\dfrac{a\sqrt{b} + b\sqrt{a}}{b\sqrt{a} - a\sqrt{b}} = \dfrac{(a\sqrt{b} + b\sqrt{a})(b\sqrt{a} + a\sqrt{b})}{(b\sqrt{a} - a\sqrt{b})(b\sqrt{a} + a\sqrt{b})} = \dfrac{ab\sqrt{ab} + a^2 b + ab^2 + ab\sqrt{ab}}{ab^2 - a^2 b}$

    $= \dfrac{a^2 b + 2ab\sqrt{ab} + ab^2}{ab(b-a)} = \dfrac{a + 2\sqrt{ab} + b}{b-a}$

7. The sum of complementary angles = $90°$

    The sum of ratio = $7 + 11 = 18$

    $\dfrac{7}{18} \times 90 = 35$ and $\dfrac{11}{18} \times 90 = 55$

    ∴ The angles are $35°$ and $55°$.

8. $5x + 4y = 8$

    $4y = -5x + 8 \Rightarrow y = -\dfrac{5}{4}x + 2$

    ∴ y-intercept = 2 and gradient = $-\dfrac{5}{4}$

    For x-intercept, $y = 0$

    Then $5x = 8 \Rightarrow x = \dfrac{8}{5}$

    ∴ x-intercept = $\dfrac{8}{5}$.

9. $\dfrac{8}{6} = \dfrac{x+y}{x} \Rightarrow 8x = 6(x+y)$

    $8x = 6x + 6y \Rightarrow 2x = 6y$

    $\dfrac{x}{y} = \dfrac{6}{2}$

    ∴ $x:y = 3:1$

10. (a) $A \cup B = \{a,b,c,d,f,g\}$
    (b) $A \cap B = \{b,d\}$
    (c) $A' = \{c,e,g,h\}$
    $A' \cap B = \{c,g\}$
    ∴ $(A' \cap B)' = \{a,b,d,e,f,h\}$

Tests for forms 1 and 2

11. $4x^2 + y = \dfrac{16x^4 - y^2}{2x + \sqrt{y}} \Rightarrow 4x^2 + y = \dfrac{(4x^2)^2 - y^2}{2x + \sqrt{y}}$

$2x + \sqrt{y} = \dfrac{(4x^2 - y)(4x^2 + y)}{4x^2 + y}$

$2x + \sqrt{y} = 4x^2 - y$ Let $\sqrt{y} = a \Rightarrow y = a^2$

$2x + a = 4x^2 - a^2$
$2x + a = (2x + a)(2x - a)$
$2x - a = 1$
$2x - 1 = \sqrt{y} \Rightarrow y = (2x - 1)^2$

---

12. $B\hat{H}C = G\hat{D}H = 50°$ (Corresponding angles)

$H\hat{B}C + E\hat{B}C = 180°$ (Angles in a straight line)

$H\hat{B}C + 110° = 180°$

$\qquad H\hat{B}C = 70°$

$H\hat{B}C + B\hat{H}C + x = 180°$

$\quad 50° + 70° + x = 180°$

$\qquad\qquad x = 180° - 120° = 60°$

$H\hat{B}C = A\hat{B}E$ (Vertically opposite angles)

But $A\hat{B}E = y$ (Alternate angles)

Since $H\hat{B}C = 70°$, then $y = 70°$

---

13. (a) $2a \times 10^2 + 3(a \times 10)^2 = 1.6 \times 10^3$

$2a \times 10^2 + 3a^2 \times 10^2 = 1.6 \times 10^3$

$2a + 3a^2 = 16$

$3a^2 + 2a - 16 = 0 \Rightarrow 3a^2 + 8a - 6a - 16 = 0$

$\qquad a(3a + 8) - 2(3a + 8) = 0$

$\qquad (a - 2)(3a + 8) = 0$

$\qquad \therefore a = 2$ or $a = -\dfrac{8}{3}$

(b) $22.5^2 - 12.5 \times 14.5 = 22.5^2 - 12.5(12.5 + 2)$
$\qquad\qquad\qquad\qquad = 22.5^2 - 12.5^2 - 12.5 \times 2$
$\qquad\qquad\qquad\qquad = (22.5 + 12.5)(22.5 - 12.5) - 12.5 \times 2$
$\qquad\qquad\qquad\qquad = 35 \times 10 - 25$
$\qquad\qquad\qquad\qquad = 350 - 25$
$\qquad\qquad\qquad\qquad = 325$

---

14. $x = B\hat{C}D$ (Alternative angles)

$y = B\hat{C}F = B\hat{C}D + z$ (Vertically opposite angles)

$y = x + z \Rightarrow x = y - z$

15. Let the numbers be x, $x+2$ and $x+4$

 Then, $x+x+2+x+4 = 84 - \frac{1}{9}(x+2)$

 $3x + 6 = 84 - \frac{1}{9}x - \frac{2}{9} \Rightarrow \frac{28}{9}x = \frac{700}{9}$

 $28x = 700 \Rightarrow x = 25$

 ∴ The numbers are 25, 27 and 29.

---

16. (a) 4, 9, 16, x, y
 $= 2^2, 3^2, 4^2$
 ∴ $x = 5^2 = 25$ and $y = 6^2 = 36$

 (b) 36, 81, 196, 441, x, y
 $= 6^2, 9^2, 14^2, 21^2$
 ∴ $x = 30^2 = 900$ and $y = 41^2 = 1681$

---

17. 6.5kg = 6.5×1000g = 6500g
 The number of packets that she got = 6500÷13
 = 500 packets.

---

18. (a) June has 30 days.
 ∴ $\frac{1}{8} \times 30$ days = $\frac{15}{4}$ days = $\frac{15}{4} \times 24$ hours = 90 hours
 = 90×60 minutes = 5400 minutes.
 (b) 1 week = 7 days = 7×24 hours = 168 hours
 = 168×60 minutes = 10080 minutes
 = 10080×605 seconds
 = 604800 seconds.

---

19. (a) $16 - 4y^2 = 4(4 - y^2) = 4(2+y)(2-y)$
 (b) $4(2+y)(2-y) = 0$
 $y = \pm 2$

---

20. (a)

21. $\frac{180°(n-2)}{n} = 140°$

 $140n = 180n - 360°$
 $40n = 360°$
 $n = 9$

 (b) $\frac{180°(n-2)}{n} = 156°$

 $180n - 360° = 156n$
 $24n = 360°$
 $n = 15$

---

22. (a) $A = \frac{h}{2}(a+b) + \frac{\pi r^2}{2}$

 $2A = h(a+b) + \pi r^2$
 $\pi r^2 = 2A - h(a+b)$
 $r = \sqrt{\frac{2A - h(a+b)}{\pi}}$

 (b) $r = \sqrt{\frac{2 \times 100 - 6(8+10)}{3.14}}$

 $r = \sqrt{\frac{200 - 108}{3.14}} = \sqrt{29.30}$

 $= 5.412$

---

23. By Pythagoras theorem, $\overline{DC}^2 = 25^2 - 15^2 \Rightarrow \overline{DC} = 20$ cm
 Then r = 20÷2 = 10 cm
 ∴ Perimeter = $\pi r + 2r + 2AD$ = (3.14×10)+(2×10)+(2×15)
 = 31.4+20+30
 = 81.4 cm
 Area = $\frac{1}{2}\pi r^2 + (AB \times AD) = \frac{1}{2} \times 3.14 \times 10^2 + 20 \times 15$
 = 157+300
 = 457 cm²

24.
P (8−3) cm Q

A

R 3 cm S

B x+0.5 cm

U 8 cm T

Total area = Area A + Area B
$$40.5 = (5 \times 2x) + [3(x+0.5)]$$
$$40.5 = 10x + 3x + 1.5$$
$$13x = 40.5 - 1.5 = 39$$
$$\therefore x = 3 \text{ cm}$$

Perimeter = PQ+QR+RS+ST+TU+UP
= 5+(2×3−(3+0.5))+3+(3+0.5)+8+2×3
= 5+6 − 3.5+3+3.5+8+6
= 28 cm

---

25. (a) $\sqrt[3]{16003.008} = \sqrt[3]{\dfrac{16003008}{1000}} = \sqrt[3]{\dfrac{2^6 \times 3^6 \times 7^3}{10^3}} = \dfrac{4 \times 9 \times 7}{10} = 25.2$

(b) $\sqrt[3]{3061.257408} = \sqrt[3]{\dfrac{3061257408}{1000000}} = \sqrt[3]{\dfrac{2^6 \times 3^3 \times 11^6}{10^6}}$

$= \dfrac{4 \times 3 \times 121}{100} = 14.52$

---

26. (a) Let $x = \dfrac{25.4 \times \sqrt{15.6}}{32.12}$

Then, $\log x = \log 25.4 + 0.5 \log 15.6 - \log 32.12$

$\log x = 1.4048 + 0.5 \times 1.1931 - 1.5068$

$\log x = 0.4946$

$x = \text{antilog } 0.4946 = 3.123 \times 10^0$

(b) Let $y = \dfrac{\sqrt[3]{82.17}}{\sqrt{15.32}}$

Then, $\log y = \dfrac{1}{3} \log 82.17 - \dfrac{1}{2} \log 15.32$

$\log y = \dfrac{1}{3} \times 1.9147 - \dfrac{1}{2} \times 1.1853 = 0.6382 - 0.5927$

$\log y = 0.0455$

$y = \text{antilog } 0.0455 = 1.11 \times 10^0$

Test 3

1. (a) $\sqrt{1.44} = \sqrt{\dfrac{144}{100}} = \dfrac{12}{10} = 1.2$

(b) $\sqrt{14} \times \sqrt{\dfrac{49}{3.5}} = \sqrt{\dfrac{49 \times 14}{3.5}} = \sqrt{49 \times 4} = 7 \times 2 = 14$

---

2. 240 = 2×2×2×2×3×5

8 = 2×2×2, 12 = 2×2×3, 16 = 2×2×2×2

∴ The number might be 5, or 2×5, or $2^2 \times 5$, or $2^3 \times 5$, or $2^4 \times 5$, or 2×3×5, or $2^2 \times 3 \times 5$, or $2^3 \times 3 \times 5$, or $2^4 \times 3 \times 5$, or 3×5.

Since, w is less than 40, therefore w = 5 or w = 2×5 = 10 or w = 3×5 = 15
or w = 2×2×5 = 20 or w = 2×3×5 = 30

∴ The possible values of w are 5, 10, 15, 20 and 30.

Tests for forms 1 and 2

3. (a) $nx = y$
$n(2) = 4 \Rightarrow n = 2$

(b) Slope, $m = \dfrac{3-0}{0-2} = -\dfrac{3}{2}$

From $y = m(x - x_1) + y_1$

$y = -\dfrac{3}{2}(x-2)$

$\Rightarrow 3x + 2y - 6 = 0$

4. (a) $\dfrac{5a \times 4b - 2ab^2 + 8a^2b \div 2a}{2ab} = \dfrac{20ab - 2ab^2 + 4ab}{2ab}$

$= 10 - b + 2 = 12 - b$

(b) $\dfrac{x^2 - y^2}{(x+y)^2} = \dfrac{(x+y)(x-y)}{(x+y)(x+y)} = \dfrac{x-y}{x+y}$

5. $3x - 2y = -3 \Rightarrow 3x - 2y = -3 /$ times $3 \Rightarrow 9x - 6y = -9$

$4x - \dfrac{3}{2}y = 3 \Rightarrow 8x - 3y = 6/$ times $2 \Rightarrow \underline{16x - 6y = 12}$

$7x = 21 \Rightarrow x = 3$

But $2y = 3 + 3x \Rightarrow 2y = 3 + 3(3) = 12 \Rightarrow y = 6$

$\therefore x = 3$ and $y = 6$.

6. $\dfrac{2\sqrt{3} - 3\sqrt{2}}{2\sqrt{3} + 3\sqrt{2}} = \dfrac{(2\sqrt{3} - 3\sqrt{2})(2\sqrt{3} - 3\sqrt{2})}{(2\sqrt{3} + 3\sqrt{2})(2\sqrt{3} - 3\sqrt{2})} = \dfrac{(2\sqrt{3})^2 - 2 \times 2\sqrt{3} \times 3\sqrt{2} + (3\sqrt{2})^2}{4 \times 3 - 9 \times 2}$

$= \dfrac{12 - 12\sqrt{6} + 18}{-6} = \dfrac{30 - 12\sqrt{6}}{-6} = 2\sqrt{6} - 5$

Then, $2\sqrt{6} - 5 = 2 \times 2.45 - 5 = 4.9 - 5 = -0.1$

7. Trapezium is a quadrilateral, therefore the sum of interior angles is 360°.
Sum of the ratio = 1+2+4+5 = 12
Then,

$\dfrac{1}{12} \times 360° = 30°, \dfrac{2}{12} \times 360° = 60°,$

$\dfrac{4}{12} \times 360° = 120°, \dfrac{5}{12} \times 360° = 150°$

$\therefore$ The degree of the angles are 30°, 60°, 120° and 150°.

8. (a)
```
    hrs   min   sec
4\   9     1     0
     2    15    15
```

(b) $\dfrac{2}{3} \times 24$ hrs = 16 hrs = 16×60×60 seconds = 57600 seconds

9. (a) $x{:}y = 2{:}5 /\times 3 \Rightarrow x{:}y = 6{:}15$
$y{:}z = 3{:}8 / \times 5 \Rightarrow y{:}z = 15{:}40$
$\therefore x{:}y{:}z = 6{:}15{:}40$

(b) $|x-2| \geq 8$
$x - 2 \geq 8$ or $-x + 2 \geq 8$
$x \geq 10$ or $-x \geq 6$
$\therefore x \geq 10$ or $x \leq -6$

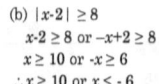

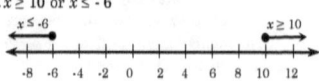

10. Given that $n(\xi) = 52$
Then, $5x - 3 + 3x + 6x - 2 + x + 2 + 2x + 3x + 2 + 3 = 52$
$$20x + 2 = 62$$
$$20x = 60 \Rightarrow x = 3$$

---

11. $\dfrac{1}{a} - \dfrac{1}{b} = \dfrac{1}{2}$

$\dfrac{b-a}{ab} = \dfrac{1}{2}$, but $b - a = c \Rightarrow b = a+c$

Then, $\dfrac{c}{a(a+c)} = \dfrac{1}{2} \Rightarrow a^2 + ac = 2c$

$$\Rightarrow a^2 + ac - 2c = 0$$

By using quadratic formula; $x = \dfrac{-B \pm \sqrt{B^2 - 4AC}}{2A}$

Where $A = 1$, $B = c$ and $C = -2c$

Then $a = \dfrac{-c \pm \sqrt{c^2 + 4 \times 2c}}{2}$

$\Rightarrow a = \dfrac{-c \pm \sqrt{c^2 + 8c}}{2}$

---

12. Let the numbers be x and $x+5$
Then, $\dfrac{x}{x+5} = \dfrac{4}{5} \Rightarrow 4(x+5) = 5x \Rightarrow x = 20$
∴ The sum of the numbers = 20+25 = 45

---

13. (a) $2a^2 - ab - b^2 = 2a^2 - 2ab + ab - b^2$
$= 2a(a-b) + b(a-b) = (2a+b)(a-b)$

(b) $9 - (x-1)^2 = 3^2 - (x-1)^2$
$= (3 + x - 1)(3 - x + 1) = (2 + x)(4 - x)$

---

14. Let the digit whose value is tens be x, then that whose value is ones is $x+6$.
Then, $x + x + 6 = 8 \Rightarrow x = 1$
∴ The number is 17.

---

15. $L^2 - l^2 = 9 \text{ cm}^2$ where $l = 4$ cm
$L^2 = 9 \text{ cm}^2 + (4 \text{ cm})^2 = 25 \text{ cm}^2$
∴ $L = 5$ cm

---

16. $\dfrac{1}{r} = \dfrac{1}{r_1} + \dfrac{1}{r_2} + \dfrac{1}{r_3} \Rightarrow \dfrac{1}{r_2} = \dfrac{1}{r} - \dfrac{1}{r_1} - \dfrac{1}{r_3} = \dfrac{r_1 r_3 - r r_3 - r r_1}{r r_1 r_3}$

∴ $r_2 = \dfrac{r r_1 r_3}{r_1 r_3 - r r_3 - r r_1}$

---

17. 10 people $\Rightarrow$ 4 days
8 people $\Rightarrow x = \dfrac{10 \times 4}{8} = 5$ days.
∴ 8 people can do the same job for 5 days.

18. (a) $\log_b a = \dfrac{\log_a a}{\log_a b}$ (By changing the base and using a as the base).

   Then, $\log_b a = \dfrac{1}{\log_a b}$ (Hence shown)

   (b) $\log_b x = \dfrac{1}{3}\log_b 8 + \dfrac{1}{5}\log_b 32$

   $\log_b x = \log_b 8^{\frac{1}{3}} + \log_b 32^{\frac{1}{5}} = \log_b 2\times 2$

   $\therefore x = 4$

19.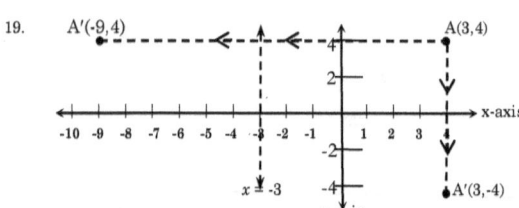

   (a) $M_x\ A(3,4) \to A'(3,-4)$
   (b) $M_{x=-3}\ A(3,4) \to A'(-9,4)$

20. $P\hat{R}S + c = 180°$ (Supplementary angles)

   $P\hat{R}S = 180° - c$

   But $a + b = P\hat{R}S$ (Ext. angle of a triangle = the sum of two opposite Int. angles)

   Then, $a + b = 180° - c \Rightarrow c = 180° - a - b$ (Hence proved)

21.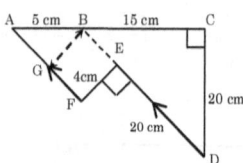

   By Pythagoras theorem, $\overline{BD}^2 = 15^2 + 20^2 \Rightarrow \overline{BD} = 25$ cm

   Again, $\overline{AG}^2 = 5^2 - 4^2 \Rightarrow \overline{AG} = 3$ cm.

   But $\overline{AF} = \overline{AG} + \overline{GF}$ and $\overline{GF} = \overline{BE} = \overline{BD} - \overline{ED}$

   $= (25 - 20)$ cm $= 5$ cm

   $\therefore \overline{AF} = 3$ cm $+ 5$ cm $= 8$ cm.

   Then, Total area = Area of triangle BCD+ Area of trapezium ABEF

   $= (\dfrac{1}{2}\times 20\times 15) + [\dfrac{4}{2}(5+8)] = 176$ cm²

   Perimeter $= (20+20+20+4+8)$ cm $= 72$ cm

22. (a) $y = C\hat{A}O + C\hat{O}A$ (ext. angle of a triangle = the sum of two opposite int. angles)

But $C\hat{O}A = 90°$

Then $C\hat{O}A = A\hat{C}O = \dfrac{180° - 90°}{2} = 45°$

$\therefore y = 45° + 90° = 135°$

(b) $\dfrac{360°}{n} = 60°$ where n is the number of sides of the polygon

$n = \dfrac{360}{60} = 6$

$\therefore$ The polygon is Hexagon.

---

23. Volume = Area of base×height

$= \dfrac{80}{360} \pi r^2 \times h = \dfrac{2}{9} \times \dfrac{22}{7} \times 7^2 \times 12 = 410.7$ cm³

But 1 litre = 1000 cm³

Then, 410.7 cm³ = $\dfrac{410.7}{1000}$ litres = 0.411 litres (3 sig. fig.)

---

24. $(5x + 75°) + 20y = 180°$ (Supplementary angles)

$5x + 20y = 105° \Rightarrow x + 4y = 21°$ ...(i)

Again, $20y = 8(6x - 5y)$ (Alternate interior angles)

$5y = 12x - 10y \Rightarrow 12x = 15y$ ...(ii)

Solve equations (i) and (ii) simultaneously and get $x = 5°$ and $y = 4°$

---

25. (a) Gradient, m = -4 and pass through point (3,0)

From $y = m(x - x_1) + y_1$

$y = -4(x-3) + 0 \Rightarrow y = -4x + 12$

$\therefore$ y-intercept is 12.

(b) $ay + (5a+2)x - 4 = 0 \Rightarrow ay = 4 - (5a+2)x$

$$y = \dfrac{4}{a} - \dfrac{(5a+2)x}{a}$$

Then, $\dfrac{-5a-2}{a} = 3 \Rightarrow a = -\dfrac{1}{4}$

y-intercept = $\dfrac{4}{a} = \dfrac{4}{-0.25} = -16$

Answers from Tests 4 – 10

Test 4
1. 60 and 180, 720
2. $y = \dfrac{5}{2}$
3. shs.11,375
4. 2 hours, 1 minute, 37 seconds
5. 1.5%
6. (a) $n = \dfrac{4}{5}$ (b) 3
7. (a) 27 oranges
   (b) 6 and 12 oranges respectively
8. (a) { } (b) {9,15,21,27,33,39}
   (c) {9,15,21,27,33,39}
9. 962.5 cm², 110 cm
18.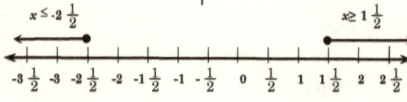
19. (a) $n = -8$ (b) $ad - bc = 0$
20. (a) $20 - 8x$
    (b) $12x - 12y - 3xy + 15y^2$
21. 238.3125 cm²
22. $x = 5$ and $y = 2$, 46 cm²

10. a:b = 11:13
11. 12 cm and 28 cm
12. (a) -1 (b) 0.33
13. 1
14. (a) (cosA+sinB)(cosA – sinB)
    (b) $\theta = 30°$
15. 125 square units
16. $x^2 + x^2 = y^2$, then $x = y\sqrt{\dfrac{1}{2}}$

    $A = x \times x = \left(y\sqrt{\dfrac{1}{2}}\right)^2 = \dfrac{1}{2}y^2$

17. (a) $6\sqrt{x+2y}$ (b) $0.5a^2$

23. (a) $6(by)^{17}$ (b) $\dfrac{16}{25x^2y^6}$
24. 146 cm²
25. (a) $n = \dfrac{t(1-a^2)}{1+a^2}$ (b) $n = -2$.

Test 5
1. 506 hrs, 47 min, and 55 s
2. (a) a = 30°, b = 60°
   (b) $\overline{BC} = \overline{BD}, \overline{AB} = \overline{BD}$ and $\overline{BD}$
   is common
   ∴ΔABD ≡ ΔCBD (SSS).
   Then, $B\hat{A}D = B\hat{C}D$
3. (a) 0.6990 (b) $\bar{2}.3980$
4. $x = 3$ and $y = -1$
5. 12 and 18 years respectively
7. $x = 2$
8. 5 years
9. $\dfrac{7}{25}$
10. 20 times
11. 2 pipes
12. (a) $\begin{pmatrix} 4 \\ -5 \end{pmatrix}$ (b) Q'(9,-9)
13. 9
14. (a) (a+b)(a+b+3)
    (b) (a – b+c)(a – b – c)
15. 2 cm

16. $5,000, $10,000 and $15,000 respectively.
17. 8 cm and 7 cm
18. £10,000
19. (a) 25, 121 (b) $\dfrac{1}{243}, \dfrac{1}{729}$
20. (b) $t_1 = \dfrac{t_2(x-L)}{c+L-x}$
21. 74 cm², 46 cm
22. (a)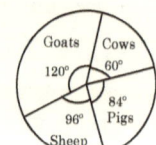

    (b) 33.33%
23. (a) $x = 4$ cm (b) 11°18'

24. (a) $x = 0.5$ or $x = 3$

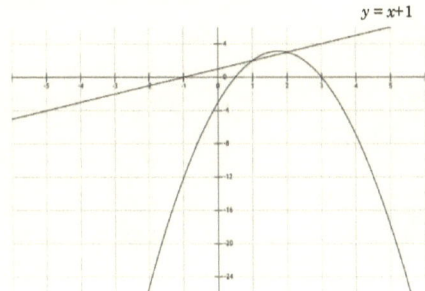

(b) $x = 1$ or $x = 2$

25. 77 cm

Test 6

1. 11 kg, 226 g, and 105 mg
2. (a) {1,3,4,5,8,10}
   (b) {1,3,4,5,7,8,9,10}
   (c) {1,2,3,4,5,6,7,8,9,10}, universal set
3. $a = 8$ if $b = 4$ and $a = 3.5$ if $b = 1$
4. 50 cm^2
5. 150 cm^2
6. $x = 3$ or $x = 9$
7. Father is 40 and his son is 20
8. 14,515,200 seconds
9. (a) $(5x+y)(x+y)$
   (b) $(a^2+b^2)(a+b)(a-b)$
10. $2x - 5$
11. 4π cm
12. $c = \dfrac{-\sqrt{ab} \pm \sqrt{ab(1-4b)}}{2}$
13. (a) -21 (b) ± 3
14. (a) 2,500 kg/m^3
    (b) 2,500,000 mm^3
15. $x = 8$
16. $x = -2$ or $x = 2.25$
17. $a = 7$ and $b = 3$
18. 6
19. $a = 5$, $b = 5\sqrt{3}$ and $c = 10$
20. 36 m^2
21. 80 cm, 300 cm^2
22. 1.49×10^0
23. $x = 15.48$ cm, $y = 17.19$ cm
24. (a) $x = \dfrac{1}{3}$ and $y = -\dfrac{5}{9}$
    (b) $\dfrac{-14 - 10\sqrt{3}}{13}$

25. (a) $x = 1.5$ (b) [Image showing triangles A'C'B' and ACB on coordinate plane with B' image reflection]
    (c) 1.125 km

Test 7

1. (a) 129,070 dm
   (b) 24 kg, and 99,991 cg
2. $x = 2$, $y = 3$ and $z = 4$
3. $x = 120°$
4. $x = 3$
5. $-5ab^2c$
6. 45 cm^2
7. $x = 7$
8. $x = 11$
9. £78,350
10. (a) $3(4b - a)(17a+2b)$
    (b) 258,925
11. (a) Finite (b) Infinite
    (c) Infinite (d) Finite
12. 60 or 120 or 180 or 240 or 360 or 720.

13. (a) $\theta = 52°$ (b) $\theta = 75°$
14. $\bar{1}.5485$
15. (a) $\dfrac{1}{12}$ (b) 0.075
16. $r = 7$ cm
17. $70°$
18. (a) -5400 (b) $n = \pm 4$
19. 96 cm^2
20. 7
21. (a) 11
    (b) If θ and α are complementary angles, then
    $\sin\theta = \cos\alpha$
    and $\cos\theta = \sin\alpha$
    $\Rightarrow \sin\theta = \cos(90° - \theta)$,
    and $\cos\theta = \sin(90° - \theta)$
    (because $\alpha + \theta = 90°$)

Then,

$$\dfrac{\sin\theta}{\cos\theta} = \dfrac{\cos(90° - \theta)}{\sin(90° - \theta)}$$

$$\Rightarrow \tan\theta = \dfrac{1}{\tan(90° - \theta)}$$

$\therefore \tan\theta \tan(90° - \theta) = 1$

22. 240 cm^2
23. 60 cm^2, 40 cm
24. (a) 15 (b) 13 – 16
    (c) 24
    (d)

Class intervals	Class mark	Frequency
9–12	10.5	3
13–16	14.5	6
17–20	18.5	5
21–24	22.5	3
25–28	26.5	7

25. 6300 metres.

Test 8

1. (a) 297 kg, 258 g, and 642mg
   (b) 11,826,000 seconds
2. $\dfrac{a}{5a+10}$
3. $\dfrac{9x+20y}{24}$
4. 6 students
5. (a) $y = 9$ (b) 1
6. 2m
7. $15°, 55°$ and $110°$
8. (a) $\overline{74}.77$ (b) $\overline{160}.1855$
9. (b) 32
10. $a = 4$ and $b = 8$
11. (a) $(a+b)(a-b)(3c-5d)$
    (b) 36,400
12. (a) 20 km (b) 1.68 cm
13. (a) -1 (b) $a = 1$
14. $y = 1.5$ or $y = 1$
15. (a) 9 m and 12 m (b) $2\sqrt{337}$ m
16. (a) $\log_b a$ (b) 1:23.5
17. $x = 6$ and $y = 2$
18. $u = -32, v = -24$ and $w = -48$
19. (a) 3 (b) $\overline{AX} = 6$ cm
20. $A''(3,-2), B''(-2,-1)$ and $C''(3,-4)$
21. (a) $\{6,9,12,15,18\}$
    (b) $\{5,7,11,13,17,19\}$
    (c) (i) $\{5,7,8,10,11,13,14,16,17,19,20\}$
        (ii) $\{8,10,14,16,20\}$
        (iii) $\{8,10,14,16,20\}$
22. (a) 3 km, 8 hm, and 7 dam
    (b) 2 m, and 05 cm
23. (a) 11.83 (b) 60.92
24. 126.8 cm^2

25. **THE BAR CHART WHICH REPRESENTS THE NUMBER OF STUDENTS IN A SCHOOL**

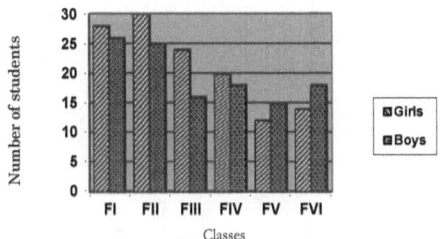

Test 9

1. (a) $y = 10$ or $y = \sqrt[5]{10^{-4}}$
   (b) $\frac{1}{2}\log\left(\frac{a+b}{4}\right)$
2. $60°$
3. Use
   $(a+b+c+d+e)+(f+g+h+i+j) = 180×5$.
   Since $a+b+c+d+e = 540°$...(i),
   $\Rightarrow f+g+h+i+j = 360°$...(ii)
   Now, use the two equations to show that
   $a+b+c+d+e = 1.5(f+g+h+i+j)$
4. 240 or 720 or 1680 or 5040
5. (a) $V = \frac{A}{6}\sqrt{\frac{A}{\pi}}$ (b) $4186.67\,cm^3$
6. (a) $a = 8$ (b) 26
7. $a = 2, b = 1$
8. 8 and 12
9. $(a-2)(a-3)(a-4)$, $a = 2$ or $a = 3$ or $a = 4$

10. $r = -\frac{h}{2} \pm \frac{\sqrt{(\pi h)^2 + 2\pi V}}{2\pi}$
11. 6.304×10^{-5}
12. $x = 3, y = 2$
13. 0.5
14. (a) $\bar{2}.9483$ (b) $\bar{1}.1590$
15. $(1,5)$
16. (a) 8 cm (b) 50.24 cm
17. (a) $(x+y)(x-y)+y$ (b) 193,915
18. (a) 5 units (b) 24
19. $r = 2$
20. (a) $\$40,000$ (b) 12.5%
21. (a) $x = 10°$
    (b) (i) 780 (ii) 480 (iii) 360
    (c) 25%
22. $1140\,m^3$
23. (a) $121\,cm^2$ (b) 66 cm
24. (a) $x = 4, y = 3$
    (c) Refer Qn.24 on test 5, $x = 4$ or $x = -1$
25. (a) (25.0, 69.6, 12.5, 23.2), 6
    (b) 35.56

Test 10

1. (a) 5 (b) 12 (c) 3
2. (a) 4 (b) -4
3. $c = 80$
4. 56 boxes of books, 40 boxes of pens and 35 boxes of rulers.
5. $\frac{-5 \pm \sqrt{817}}{36}$
6. $(3^{10b-2})(5^{2a-3b})$
7. $a = 8$ and $b = 2$

8.

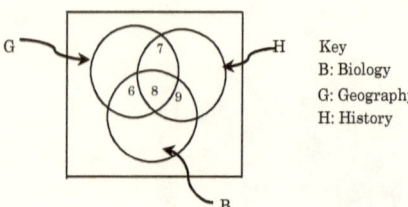

(a) 14   (b) 7   (c) 30

9. shs.1980, 19.8%
10. 3
11. (a) $x = 0, y = 1$ (b) 300
13. 115°

12. $r = \dfrac{-R\pi h \pm \sqrt{\pi^2 h^2 - 4\pi h(R^2\pi h - 3V)}}{2\pi h}$

14. 74 kg
15. (a) 50

16. (a)

$-3.5 \leq x \leq 0.5$

```
●——|——|——|——|——|——|——|——●
-3.5 -3 -2.5 -2 -1.5 -1 -0.5 0 0.5
```

(b) 4 or 0.25

17. (a) $\dfrac{206}{2475}$ (b) 17.8

18. 13:7 respectively.
19. £84,000
20. 3.5 cm
21. (a)

Class mark	Real limits	Tally mark	Frequency	C. frequency
5	0–10	/	1	1
15	10–20	////	5	6
25	20–30	//// /	6	12
35	30–40	//// //// //	12	24
45	40–50	//// ///	8	32
55	50–60	//// /	6	38
65	60–70	//	2	40

(b)

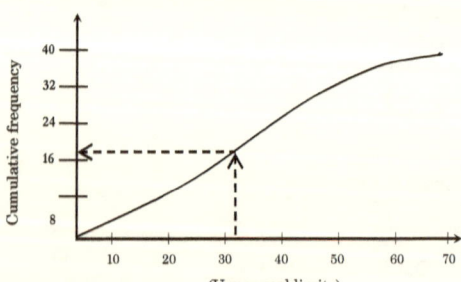

(Upper real limits)

∴The number of students who got at least 45 is 18.

22. (a) 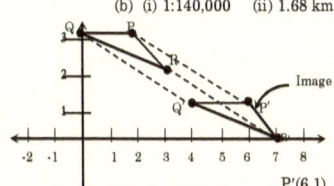 (b) (i) 1:140,000 (ii) 1.68 km

   P'(6,1), Q'(4,1) and R'(7,0)

23. 76.51 cm

24. (a) $x = \dfrac{5 \pm \sqrt{105}}{4}$ (b) $l = \dfrac{A - \pi r^2}{\pi r}$, 50

25. (a) (i) $1.87\dot{7}\dot{8}$ (ii) $0.21\dot{3}\dot{2}$ (b) $\dfrac{2111}{18591}$

PART 2

Tests for Forms 3 And 4

Test 1

Section A

1. Find the perimeter and area of a ten-sided regular polygon inscribed in a circle of radius 10 cm.

2. Given that vector **a** = 3i – 2j and **b** = 3i+7j. Find
   (a) **c**, if **c** = **a**+**b**   (b) the modulus of 2**a** – **b**

3. If $x = 0.3\dot{5} - 0.\dot{2}\dot{3}$, express the value of x in the form of $\dfrac{a}{b}$ where $b \neq 0$.

4. A ship left the point (25°N, 60°E) and sailed due east for one day, to the point (25°N, 90°E). Calculate the average speed of the ship in km/h.

5. Given that a^2 is directly proportional to b and is inversely proportional to the square root of c. If $a = 2$ when $b = 3$ and $c = 4$, find b when $a = 4$ and $c = 9$.

6. 5 and 1215 are the first and n^{th} terms of a geometric progression respectively. If the sum of the first n terms is 1820, find the common ratio.

7. (a) Round off each of the following numbers to one decimal place.

    $a = 22.372, b = 44.835$

   (b) Use the result obtained in (a) above to evaluate

    $$\dfrac{a \times b}{a^2}$$

8. The number of diagonals of n-sided polygon is given by the formula $\dfrac{1}{2}n(n-3)$. Write the names of the polygon which have the following number of diagonals:
   (a) 27
   (b) 5

9. Use matrices to solve the following simultaneous equations:
    $$2x + 2y = 10 \text{ and } 2x - 3y = 5$$

10. The expression $4x^2 - 25x + px + 25$ has equal roots, find the possible values of p.

11. (a) Factorize completely the expression $\sin^4 a - \cos^4 a$.
    (b) Given that $\sin A = 0.6$ and that $A+B = 45°$. Without using any mathematical aids, find the value of
        (i) $\sin B$ (ii) $\cos B$ (iii) $\tan B$

12. Find the value of a, b and c if $2a+9b = 4a+8c = 4$ and $\dfrac{6b-c}{7} = \dfrac{1}{4}$.

13. Given that $\log a^3 b = 3.5$ and $\log a^2 \cdot \sqrt{b} = 1.5$, evaluate
    (a) $\log(ab)^3$
    (b) $\log \sqrt{\dfrac{b}{a}}$

14. The terms a^{-2}, b^{-2} and c^{-2} are consecutive terms of a geometric progression, show that a, b and c are also consecutive terms of GP. Hence show that $\log a, \log b$ and $\log c$ are consecutive terms of an arithmetic progression.

15. Use mathematical tables to evaluate the following:
    (a) $\dfrac{\log_7 15}{\log_3 8}$
    (b) $\dfrac{\sqrt[3]{0.0122} \times 10.543}{(11.2)^2 \times 0.00123}$

16. Find the value of a, b and c such that $\begin{pmatrix} a & 2b-3 \\ 5 & c-b \end{pmatrix} = \begin{pmatrix} 4 & 9 \\ 5 & 3 \end{pmatrix}$

17. (a) What are the next two terms in sequence below?
    $$\dfrac{1}{2}, \dfrac{3}{5}, \dfrac{6}{11}, \dfrac{12}{23}, \ldots$$
    (b) Find the general term of the sequence
    $$\dfrac{13}{9}, \dfrac{12}{13}, \dfrac{11}{17}, \dfrac{10}{21}, \ldots$$

18. A straight line L has a y-intercept of 2 and x-intercept of -2. What is
    (a) the gradient of L?
    (b) the equation of L?
    (c) the equation of a line passing through point (3,3) and perpendicular L?

19. Given that n(X) = 7, n(Y) = 8, n(X∪Y)' = 5 and n(μ) = 17 where μ is a universal set. Find
    (a) n(X∪Y),
    (b) n(X∩Y), and
    (c) n(X')

20. If a letter is chosen randomly from the word 'EQUATION', what is the probability that it is
    (a) a vowel?
    (b) neither letter Q nor N?

Section B

21. Given a function $f(x) = -2x^2 + 4x - 3$, find
    (a) the turning point of $f(x)$,
    (b) the maximum or minimum value,
    (c) the symmetry line, and
    (d) the range of $f(x)$.

22. Find the value of y from the figure 2.1.22 below.

    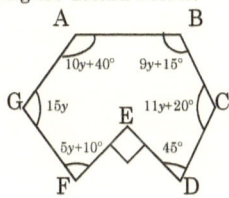

    Figure 2.1.22.

23. The following data shows the scores of forty students in a certain mathematics examination.

62	35	32	30	54	67	29	79	45	48
58	62	35	84	72	55	73	59	60	57
47	67	70	41	32	45	47	50	73	28
25	59	39	49	50	64	88	64	54	52

    (a) Group the scores in class intervals 20–30, 30–40, 40–50, etc. and construct the frequency distribution table for the data.
    (b) Draw a cumulative frequency curve and use it to estimate the median.
    (c) Use an assumed mean as 55, to calculate the mean score.

24. (a) If the matrix $\begin{pmatrix} x+2 & 7 \\ 4x-2 & 3x+5 \end{pmatrix}$ is a singular matrix, what are the possible values of x?

(b) Matrix $A = \begin{pmatrix} 2 & 3 \\ 4 & 8 \end{pmatrix}$ and $B = \begin{pmatrix} 0 & -1 \\ 2 & -3 \end{pmatrix}$. Find
  (i) the matrix AB.
  (ii) the matrix A^{-1}.
  (iii) $2(A^{-1}+A) - B$.

25. Mr. Juma commenced a business on 1st January, 2007 with a capital of $260,000.

  Jan 2: purchased goods for cash $70,000
      3: sold goods for cash $90,000
      7: purchased goods for cash $30,000
     13: bought packing materials $5,000
     19: paid transport charges $3,500
     21: bought more goods for cash $20,000
     24: paid insurance for cash $6,000
     27: sold goods for cash $60,000

(a) Enter the above transactions in cash and sales accounts and balance them at the end of the month.

(b) Prepare the trial balance.

Tests for forms 3 and 4

Test 2

Section A

1. (a) If $f(x) = \cos x$, find the range of $f(x)$.
   (b) Show that $(\sin x - \cos x)^2 + (\sin x + \cos x)^2 = 2$.

2. The fifth term of an arithmetic progression is 31 and the sum of the first twelve terms is 400. Find the first term and the common difference of the progression.

3. Given that $n(X \cap Y') = 15$, $n(X' \cap Y) = 20$ and $n(X' \cap Y') = 4$ when $n(\mu) = 50$ where μ is a universal set.
   (a) Represent the information on the Venn diagram.
   (b) Find
      (i) $n(X)$
      (ii) $n(Y)$
      (iii) $n(X \cap Y)$
      (iv) $n(X' \cup Y)$

4. The arithmetic and geometric means of two numbers are 10 and 8 respectively. Find the possible values of the numbers.

5. Use mathematical tables to evaluate the following and write the answer into two decimal places.
$$\frac{(25.32)^2 \times \sqrt{45.38}}{(5.834)^3 \times \sqrt[3]{84.12}}$$

6. The area of a regular hexagon which is inscribed in a circle is 1175.6 cm². Calculate the area and circumference of the circle.

7. If $f(x) = \dfrac{5}{x-a} + \dfrac{6+a}{x+2a}$ and $f(2)$ is undefined, calculate the value of a and $f(4)$.

8. Given the points A(0.7,-1), B(-0.8,-0.2) and C(-0.2,0.2). Find
   (a) the length of the line segments (i) AB (ii) AC
   (b) the midpoints of the line segments (i) BC (ii) AC
   (c) the equations of the lines
      (i) BC, and
      (ii) AC in the form of $\dfrac{x}{a} + \dfrac{y}{b} = 1$ where a and b are any real numbers.

9. In what proportions of two types of juice containing 25% and 35% of sugar should be mixed in order to get the mixture containing 28% of sugar?

10. (a) Given that **r** = 2**i** − 4**j**, **s** = 4**i**+**j** and **t** = 2**r** − **s**. Calculate the value of |**t**|.
    (b) Find the equation of the line passing through point (-1,2), and perpendicular to the line $2x-3y+5 = 0$. Write the answer in the form of $ax+by+c = 0$ where a, b and c are integers.

11. A and B are the points at (45°N, 86.5°E) and (45°N, 93.5°W) respectively. Find
    (a) the time at B if the time at A is 2:28 p.m.
    (b) the distance from A to B along
        (i) the great circle.
        (ii) the parallel of latitudes.

12. Multiply the mass (15 kg, 150 g, and 150 mg) by 23.5

13. (a) Find the numbers in the ratio 2:3:4:5 whose sum is 21.
    (b) If $a:b:c:d = e:f:g:h$, show that $\dfrac{a+b+c+d}{e+f+g+h} = \dfrac{a}{e}$

14. If α and β are the roots of a quadratic equation $ax^2+bx+c = 0$. Show that
    (a) $\alpha+\beta = -\dfrac{b}{a}$, and
    (b) $\alpha\beta = \dfrac{c}{a}$
    Hence show that $ax^2 - a(\alpha+\beta)x+a\alpha\beta = ax^2+bx+c = 0$

15. (a) From the equation $ax^4+bx^2+c = 0$, show that
    $$x = \sqrt{\dfrac{-b \pm \sqrt{b^2 - 4ac}}{2a}}$$
    (b) Use the formula in (a) above to solve the equation
    $$x^4 - 13x^2+36 = 0$$

16. From the figure 2.2.16 below, prove that $A\hat{B}D = \dfrac{1}{2}a - b$

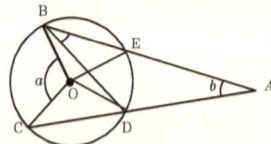

Figure 2.2.16.

Note: O is the centre of the circle BCDE.

17. A square and a circle have the same areas. Prove that the circumference of the circle is smaller than the perimeter of the square.

18. Find the inverse of each of the following matrices:

   (a) $\begin{pmatrix} 2 & 2 \\ 4 & 5 \end{pmatrix}$

   (b) $\begin{pmatrix} 2 & -7 \\ -3 & 8 \end{pmatrix}$

19. Triangle XYZ is such that XY = XZ and XY:YZ = 5:6. If the area of this triangle is 108 cm², find its perimeter.

20. Given that a varies directly as (b^2+2) and inversely as the square root of $(c+1)$, and that $a = 3$ when $b = 2$ and $c = 3$. Find a when $b = 4$ and $c = 3$

    (The constant of proportionality is a positive number)

Section B

21. Calculate the surface area of the following triangular prism.

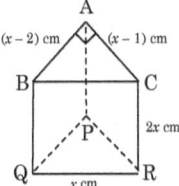

Figure 2.2.21.

22. If $f(x) = \begin{cases} -2 & \text{when } x \geq 2 \\ x+3 & \text{when } -3 \leq x < 2 \\ 3 & \text{when } x < -4 \end{cases}$

    (a) Sketch the graph of $f(x)$
    (b) State the domain and range of $f(x)$
    (c) Find
       (i) $f(3)$ (ii) $f(-3.25)$

23. (a) Minimize $F(x,y) = 4x+5y$ subject to
         $x \geq 0, y \geq 0, 3x+2y \leq 10$ and $x+y \geq 4$

    (b) Mrs. Ali wants to buy two types of gowns, type A and B at shs.10,000 and shs.5,000 each respectively. She has a total of shs.100,000 and she want to buy not more than 14 gowns altogether. When she sell them, her profit on each type A is shs.4,500 and on each type B is shs.2,500. How many of each type of gown she has buy for the maximum profit?

24. (a) A bag contains 4 oranges and 6 mangoes. If a fruit is picked at random from the bag and then replaced before the second fruit is picked, find the probability of picking
    (i) two oranges (ii) one orange and one mango.
  (b) The probability that Ali can pass the exam is 0.65 and the probability that Jane can fail the exam is 0.43. Find the probability that
    (i) both Ali and Jane can pass the exam.
    (ii) either Ali or Jane can fail the exam.

25. The frequency distribution table below shows the weights of villagers in a certain village.

Weight (in kg)	1–20	21–40	41–60	61–80	81–100	101–120
Frequency	18	33	36	60	45	8

Calculate (a) the mean weight,
          (b) the median weight, and
          (c) the modal weight

Test 3

Section A

1. Given that the quadratic equation $ax^2+(4-6a)x+25 = 0$ has equal roots.
   (a) Find the possible value of a
   (b) Write the equation.

2. Find the length of the chord of a circle of radius 7cm if the measure of the arc joining two end points of that chord is
   (a) 40°
   (b) $\frac{2\pi}{3}$ radians
   (c) 2.8 radians

3. From the following circle, AB is a straight line. Prove that $y = 180°+\theta+\beta - \alpha$ if O is the centre of the circle.

   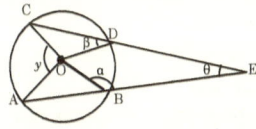

   Figure 2.3.3.

4. (a) One end-point of the line segment whose midpoint is (3,-5), is (4,-7). Find the other point of the line segment.
   (b) Solve the equation $(x-2)^4+2x-x^2 = 4-2x$

5. When the time at point A(25°S, 160°E) on the earth's surface is 2:30 am, the time at point B is 7:00 am. What is the longitude of point B?

6. The angle of elevation of the top of the tower 30 m high is observed from the top and the bottom of a tree and found to be 30° and 60° respectively. Calculate
   (a) the height of the tree.
   (b) the distance from the tree to the tower.

7. The expression $ax^3+bx^2 - 4x+2$ has a remainder 3 when divided by $(x-1)$. When it is divided by $(x-2)$, it leaves the remainder 22. Find the possible values of a and b.

8. An arithmetic progression of $(b-2)$ terms has a and z as the first and the last terms respectively. Find in terms of a, b and z,
   (a) the tenth term,
   (b) the term before the last term, and
   (c) the sum of the first $(b-3)$ terms.

Tests for forms 3 and 4

9. A cyclic quadrilateral is drawn inside the circle as shown below, given that the length $AB = 5$ cm, $BC = 6$ cm, $CD = 8$ cm, and $\angle BAD$ is a right angle. Find the area of the shaded part.

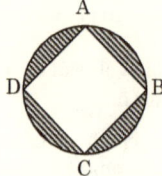

Figure 2.3.9.

10. (a) The circle $(x - a)^2 + (y - b)^2 = r^2$ passed through points $(14, 12)$, $(11, 15)$ and $(-10, -6)$. Find the values of a, b and r.
    (b) Evaluate x into three significant figures if $8^x = 24$

11. Given that **a** $= 3\mathbf{i} - 2\mathbf{j}$, **b** $= \mathbf{i} + 3\mathbf{j}$ and **c** $= 2\mathbf{i} - \mathbf{j}$.
    Find the following:
    (a) $2\mathbf{a} + 3\mathbf{b} - 4\mathbf{c}$
    (b) $|4\mathbf{c} - 2\mathbf{a} - 3\mathbf{b}|$

12. Given that, the roots of the quadratic equation $2x^2 - 3x + 4 = 0$ are α and β. Find the quadratic equation whose roots are 3α and 3β.

13. In a certain village of 25 men, 14 among them are fishermen and 16 are farmers. Each one must work at least one job between the two. What percentage of men are both farmers and fishermen in the village?

14. Describe the matrix of reflection along the y-axis, and then find the image of point $(3, 5)$ after reflected along this axis.

15. (a) Solve for x from equation $3^x + 2(3^{-x}) = 3$
    (b) Solve the inequality $\dfrac{2x + 4}{x - 2} \geq 4$

16. The sum of five consecutive terms of an arithmetic progression is 0 and the common difference is 3. Find the sum of the next five consecutive terms of this AP.

17. Find the interest obtained on $70,000 for three years at 3.5% compounded annually.

18. Use mathematical tables to find the value of a when $\dfrac{1}{a} = (3.52)^2 - (3.25)^2$, and write the answer in the form of $a \times 10^n$ where $1 \leq a < 10$ and n is any integer, into three significant figures.

19. Use mathematical tables to evaluate
    (b) tan(215°)   (b) sin(316°)   (c) cos(200°)

20. Two sides of a regular pentagon are produced to meet at point A. Calculate the size of angle A.

Section B

21. Given that $f(x)=\begin{cases} -2 & if \quad x \geq 4 \\ x-1 & if \quad x \leq -1 \\ -1 & if \quad -1 < x < 4 \end{cases}$

    (a) Sketch the graph of $f(x)$.
    (b) Find f(-2).
    (c) State the domain and range.
    (d) Is f(x) one-to-one function?

22. The frequency distribution table below shows the ages of the people in a certain place.

Age in years	12–19	20–27	28–35	36–43	44–51
Number of people	9	15	8	12	8

    Calculate (a) the mean,
             (b) the median, and
             (c) the mode.

23. The figure 2.3.23 below is a rectangular block with sides AB = 8 cm, AP = 10 cm, and QR = 6 cm.

    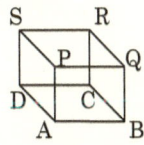

    Figure 2.3.23.

    Calculate

    (a) the length AC,
    (b) the length AR,
    (c) the angle that AR makes with the base ABCD, and
    (d) the angle that plane ABRS makes with the base ABCD.

Tests for forms 3 and 4

24. (a) Two ordinary dice were rolled simultaneously. Find the probability of turning up two numbers whose sum is
    (i) less than 5.
    (ii) greater than 8.
    (iii) greater than or equal to 6.
    (b) A coin is tossed and an ordinary die is rolled. Find the probability that a head appears on the coin and an even number appears on the die.

25. On 31st January 2005, the double entry transaction of a certain company was as follows:-

    DR   Cash A/c   CR

Date	Particulars	Amount (shs.)	Date	Particulars	Amount (shs.)
Jan, 1	Capital	8,000,000	Jan, 2	Purchases	2,000,000
4	Sales	2,500,000	9	Purchases	1,500,000
10	Sales	1,800,000	15	Furniture	800,000
22	Sales	3,200,000	16	Motor vehicles	1,500,000
			18	Purchases	1,080,000
			20	Salaries	900,000

(a) From the transactions above, balance and bring down the balance at the end of the month of January.

(b) Prepare the trading and profit and loss accounts
Note: Closing stock shs.780,000

(c) Construct the balance sheet.

Tests for forms 3 and 4

Test 4

Section A

1. Use the substitution $a = \dfrac{1}{x}$ and $b = -\dfrac{1}{y}$ to solve for x and y from the following simultaneous equations.

$$\begin{cases} \dfrac{3}{x} - \dfrac{2}{y} = 12 \\ \dfrac{1}{2x^2} - \dfrac{2}{3y^2} = -4 \end{cases}$$

2. Calculate the surface area of the following hemisphere of radius 35 cm.

    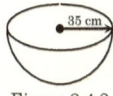

    Figure 2.4.2

3. If $a = \sqrt{b}$, show that

$$\dfrac{\sqrt{a} + 2b^{\frac{1}{4}}}{2a - \sqrt{b}} = 3 \times \sqrt[4]{\dfrac{1}{b}}$$

4. Without using mathematical tables, evaluate

$$\dfrac{\log_2 12.8 + \log_2 5}{\log_3 24.3 + \log_3 90}$$

5. Write the function $f(x) = -2x^2+3x+3$ in the form of $f(x) = a(x+b)^2+c$ where a, b and c are real numbers, hence determine
    (a) the turning point,
    (b) the axis of symmetry,
    (c) the maximum or minimum value, and
    (d) the range of $f(x)$.

6. A line whose slope is 4 and passes through point (1,-2), intersects a line $ax+2y = 24$ at point (3,b). Find the value of a and b.

7. (a) If the volume, V, of a cylinder is given by $V = \pi r^2 h$ and its surface area, A, is given by $A = 2\pi r^2 + 2\pi rh$, express V in terms of A, π and h only.

    (b) Find the value of V in (a) above if $A = 396$ cm², $\pi = \dfrac{22}{7}$ and $h = 2$ cm.

Tests for forms 3 and 4

8. John bought some packets of sweets for shs.1500. If the price of each packet had been reduced by shs.50, he would receive 5 more packets for his money. How many packets did he buy?

9. Find the values of a and b if the expression x^3+ax^2+bx-4 is exactly divisible by x^2-4. Hence, find the other factor.

10. Given that,
    $1 = tzs.2,280,
    kes.1 = tzs.23,
    ugx.1 = tzs.0.61.
    Change
    (a) $25.3 into tzs.
    (b) tzs.20,000 into kes.
    (c) kes.5.25 into ugx.

11. (a) Find the sum of of the first three-hundred integers, which are not divisible by 4.
    (b) How many terms of an arithmetic series 7+11+15... must be taken in order to get the sum equals to 525?

12. Calculate the area of shaded region from the circle below, if the radius is 5 cm and O is the centre.

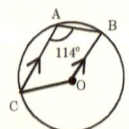

Figure 2.4.12.

13. From the circle ABC below, AC and OB are parallel, and O is the centre of the circle. Find the angles
    (a) $B\hat{O}C$ (obtuse angle),
    (b) $A\hat{B}O$, and
    (c) $A\hat{C}O$

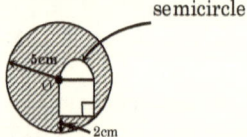

Figure 2.4.13.

14. (a) The point P(2,4) is reflected to P'(-3,5). Find the equation of the line of reflection.
    (b) Evaluate $\left(\dfrac{8}{27}\right)^{\frac{2}{3}} + \left(\dfrac{81}{25}\right)^{-\frac{1}{2}} - \sqrt{\left(\dfrac{0.81}{1600}\right)^{-\frac{1}{2}}}$

15. (a) Write the ratio 25.8:77.4 in the form of 1:n where n is a real number.
    (b) On the circle below, O is the centre. Prove that, ΔATO ≡ ΔBTO and hence show that, AT = BT.

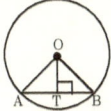

Figure 2.4.15

16. Given that **a** = 3*i*+5*j* and **b** = 5*i* − *j*. Find
    (a) 2**a**+**b**,
    (b) the magnitude of 2**a**+**b**,
    (c) the direction of vector 2**a**+**b** that makes with positive x-axis, and
    (d) the direction cosines of vector 2**a**+**b**.

17. The relation R = {(x,y): $y \geq x^2$ and $y < 6 - |x|$}
    (a) Draw the graph of R.
    (b) State the domain and range from the graph drawn in (a) above.

18. (a) Find the equation of a line which passes through points A(3,5) and B(4,7).
    (b) Find the equation of a line parallel to the equation of a line obtained in (a) above and passes through the point C(7,5).

19. From figure 2.4.19 below, AP = 0.5x, CP = x, AB = 15 cm, PQ = 2x and CQ = 8 cm.

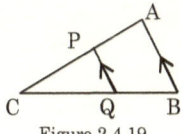

Figure 2.4.19.

Find
(a) the value of x,
(b) the length of QB, and
(c) the perimeter of triangle ABC.

20. (a) Given that the universal set ξ = {Natural numbers less than 20}, and its subsets A = {Odd numbers or multiples of 3} and B = {Even numbers}.

    Describe the following sets in words:
    (i) A∪B
    (ii) A∩B

    Find n(A∩B)'

(b) Reflect the point P(-3,4) along the line $y+3 = 0$

Section B

21. A tailor has 16 m² and 17 m² of type A cloth and type B cloth respectively. He needs to make skirts and blouses by using 2 m² and 1 m² of type A cloth and using 1m² and 2m² of type B cloth respectively. If the price of skirt is shs.5,000 each and that of blouse is shs.4,500 each, how many skirts and blouses he has to make in order to get the maximum profit?

22. A person commenced a business on 1st April, 2001 with a capital and opening stock of $20,000 and $2,000 respectively.

    April,3: purchased goods for cash $8,000
        5: sold goods for cash $10,000
        6: paid wages $1,000
       13: purchased goods $10,000
       17: sold goods $15,000
       25: paid salaries $4,000
          and paid insurance $1,500
       28: cash sales $5,000

    The closing stock in hand on 30th April, 2001 was $1200.

(a) Open the cash account and bring down the balance as on 30th April.

(b) Prepare the final accounts and construct the balance sheet as on 30th April.

23. From figure 2.4.23 below, calculate
    (a) the total surface area (b) the volume

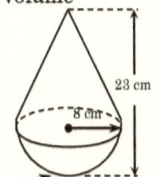

Hemisphere of radius 8 cm
Figure 2.4.23

24. The distribution table below represents the masses of 80 people in Mkoroshoni village.

Mass in kg	1–10	11–30	31–40	41–60	61–70	71–80
Number of people	3	7	$3x+1$	29	20	$10-x$

   (a) Find the value of x.
   (b) Draw the histogram and the frequency polygon on the same axis to represent the data.
   (c) Estimate the mean mass.

25. The function f, of x, is defined as $f(x) = 2x - [x]$
    (a) Draw the graph of $f(x)$
    (b) State the domain and range of f.

Test 5

Section A

1. (a) Given that $\tan(\theta - \alpha) = 2$ and $\tan \alpha = 3$. Find the value of $\tan \theta$.

   (b) Solve the equation $2\sin^2 \theta + \cos \theta = 2$ for which
   $$0 \leq \theta \leq 2\pi.$$

2. From the figure 2.5.2 below, calculate the area of the minor segment PQ if the radius of the circle is 14 cm and O is the centre of the circle.

   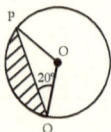

   Figure 2.5.2.

3. A jet is flying on a course of 300° with an air speed of 250 km/h. The wind is blowing from 029° at 42 km/h. Calculate
   (a) the ground speed of the jet,
   (b) the track, and
   (c) the drift.

4. (a) Solve the logarithmic equation below.
   $$\log_2(x+2) + \log_2(x-5) = 3$$

   (b) If $\log a = 0.2$ and $\log b = 0.4$, evaluate

   (i) $\log \dfrac{a^2}{\sqrt{b}}$

   (ii) $\log \sqrt{\dfrac{ab}{\sqrt[3]{a}}}$

5. The perimeter and area of a triangle whose one of its sides is 10 cm are 44 cm and $8\sqrt{66}$ cm². Find the lengths of the other two sides, and hence show that the triangle is an obtuse-angled triangle.

6. (a) Approximate a number 12.5701 into
    (i) one decimal place.
    (ii) five significant figures.
   (b) Given that $\sqrt{2} = 1.414$, without using mathematical tables or a calculator, evaluate

   $$\frac{2+\sqrt{2}}{2-\sqrt{2}}$$

7. Construct a triangle ABC such that $\angle A = 60°, \angle B = 90°$ and $\overline{AB} = 6$ cm without using a protractor.

8. Find the sum of the following series:
   (a) 21+19.5+18+...-7.5
   (b) 8+10+12.5+... (up to the tenth term)

9. The consecutive non zero terms of a progression are $x - 6$, $x+2$ and $3x+6$. Find those terms if the progression is
   (a) arithmetic progression.
   (b) geometric progression.

10. Consider the following methodology.
    For any integers, x and y, given that, $x = 0.\dot{y}...(i)$

    Then $10x = y.\dot{y} = y + 0.\dot{y}...(ii)$

    $10x = y + x \quad (x = 0.\dot{y})$

    $9x = y$

    $\therefore x = \dfrac{y}{9}$

    Now, use this methodology to convert the following recurring decimals into fractions:

    (a) $0.\dot{7}$
    (b) $0.2\dot{3}\dot{5}$
    (c) $1.0\dot{3}\dot{2}$

11. Point Q is 80 km east of P, and R is 120 km on a bearing of 150° from Q. Calculate
    (a) the distance from P to R, and
    (b) the bearing from P to R.

12. In a certain school, 60% of pupils are girls. If 75% of the boys and 25% of the girls study science subjects, what is percentage of the pupils who do not study science subjects?

13. Use matrices to solve the simultaneous equations
$$\begin{cases} \dfrac{1}{2}x - 3y = -1 \\ \dfrac{1}{5}x + \dfrac{1}{2}y = 3 \end{cases}$$

14. (a) If a*b is defined as $a^2 - b^2$, find the value of
    (i) 2*-3 (ii) (-1*4)*(2*3) (iii) x if $x*3 = 16$
    (b) A and B are the sets defined as, A = {x: x > -4} and B = {x: $x \leq 8$} where x is a real number. Find the set A∩B.

15. Use the balance sheet equation to find X, Y and Z in the table below.

ASSETS		CAPITAL	LIABILITIES	
Fixed	Current		Long term	Short term
7,000	8,500	10,000	X	1,320
2Y	12,000	10,000	8,000	Y
160,000	0.5Z	1.5Z	0.25Z	85,000

16. (a) Write 0.078026 correct to
    (i) three decimal places.
    (ii) three significant figures.
    (b) How many significant figures does the number 0.08340 have?
    (c) What is the place value of 6 on the number 10.3564?

17. Pipes A and B can fill the tank for 3 hours and 4.5 hours respectively. How much time will be taken if the pipes A and B run together to fill the tank?

18. If A and P are the respective area and perimeter of n-sided regular polygon inscribed in a circle of radius r, prove that
$$A = \frac{P.r}{2}.\cos\left(\frac{180°}{n}\right)$$

19. (a) Find the matrix X from the equation
$$\begin{pmatrix} 2 & 8 \\ 3 & 4 \end{pmatrix} - \frac{1}{2}X = \begin{pmatrix} 4 & -2 \\ 3 & 1 \end{pmatrix}^2$$

(b) Given that, the matrix $A = \begin{pmatrix} 5 & 5 \\ -5 & -5 \end{pmatrix}$

  (i) Find A^2
  (ii) What can you say about the matrix A^2?

20. Consider the following arrow diagram representing a certain relation.

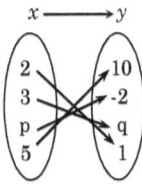

Figure 2.5.20.

If the relation shown on the figure above is $y = ax+b$ where a and b are real numbers, find

(a) the value of a and b,

(b) the value of p and q, and

(c) the inverse of the relation in the equation form.

Section B

21. A farmer decided to sell his oranges and mangoes for tzs.100 and tzs.120 each respectively. He wants to get at least tzs.2,840 after selling these two types of fruits. He also wants to sell not more than 15 oranges and 12 mangoes.
    (a) How many of each should he sell so as to
        (i) minimize the profit?
        (ii) maximize the profit?
    (b) What the amount of money will he get in a(i) and a(ii) above?

22. (a) The image of point (x,y) under the transformation T is (x',y') such that,
$$\begin{pmatrix} x' \\ y' \end{pmatrix} = \begin{pmatrix} 3 & 2 \\ 0 & -1 \end{pmatrix} \begin{pmatrix} x \\ y \end{pmatrix} + \begin{pmatrix} 5 \\ 3 \end{pmatrix}$$

    (i) Find the coordinates of the image of point (3,5) under the transformation T.
    (ii) At what point will the transformation T map the image at point (5,3)?

(b) Find the image of point (-2,4) after the rotation of 60° clockwise direction about the origin.

(c) The matrices P, R and X are defined as $\begin{pmatrix} 2 & 3 \\ -1 & 0 \end{pmatrix}$, $\begin{pmatrix} 3 & 5 \\ 2 & 7 \end{pmatrix}$ and $\begin{pmatrix} \frac{2}{15} & -\frac{7}{15} \\ -\frac{1}{30} & \frac{11}{30} \end{pmatrix}$ respectively. It is given that $(PQ+R)^{-1} = X$.

Find the matrix Q.

23. (a) A function f is defined as $f(x) = x^2+1$.
    (i) Find the domain and range of f.
    (ii) For what value(s) of x will the function f yield the value equals to 5?

   (b) Given that $f(x) = \begin{cases} -2 & if \quad x > 2 \\ 0 & if -2 < x \leq 2 \\ 2 & if \quad x \leq -2 \end{cases}$

    (i) Sketch the graph of $f(x)$.
    (ii) Find $f(5)$ and $f(-2)$.
    (iii) State the domain and range.
    (iv) Is f one-to-one function? why?

24. The frequency distribution table below shows the scores of 100 students in mathematics examination.

Class intervals	1–10	11–20	21–30	31–40	41–50	51–60	61–70
Frequency	9	2x+9	18	4x	16	10	2x-2

   (a) Find the value of x.
   (b) Calculate the mean mark.
   (c) Draw the histogram and use it to estimate the modal mark.

25. (a) A bag contains three white balls, four blue balls and five red balls. Two balls are drawn from the bag at random one after another without replacement. What is the probability of drawing two balls,
    (i) both being white.
    (ii) both being of the same colour.
    (iii) one being white and the other being red.

   (b) A two digit number can be formed by using the digits 1, 2, 3, 4 and 5.
    (i) How many numbers can be formed if the digits in each number are all different?
    (ii) How many numbers in b(i) are greater than 300?
    (iii) How many numbers in b(i) are even numbers?
    (iv) Find the probability of forming the numbers in b(ii) then in b(iii) above.

Test 6

Section A

1. The universal set $U = \{x \in \mathcal{R}: -3 \le x < 10\}$ and its subsets A and B are $\{x: x > 0\}$ and $\{x: x \le 5\}$ respectively. Represent the following sets on the number line:
   (a) A
   (b) B
   (c) A∪B
   (d) A∩B

2. Simplify the following without using mathematical tables.
$$\frac{\sin 115° \cos 135° - \tan 150° \cos 335°}{\cos 25° \sin 300°}$$

3. The equation $4x^2+(2a+10)x+5a = 0$ has equal roots. Find
   (a) the value of a, and
   (b) the solution of the equation.

4. (a) Solve for x from the equation $3^{2x+1} = 6^{5-x}$.
   (b) Juma has ten notes of shs.500 and shs.2,000 altogether, making a sum of shs.8,000. How many notes of shs.500 and that of shs.2,000 does he have?

5. Without using table of values, sketch the graph of $y = x^2+2x-3$, and hence use it to estimate the values of x from the equation $x^2+2x - 6 = 0$.

6. The ratio of the areas of the larger and smaller circles is 9:25. If the circumference of the larger one is 20 cm, find
   (a) the ratio of their radii and the radius of the smaller circle.
   (b) the area of the smaller circle and that of the larger one.

7. (a) Given that $x^2+x^{-2} = 18$, find the value of $x - x^{-1}$.
   (b) If α and β are the roots of the equation $4x - 5x^2+7 = 0$, form the equation whose roots are $α^2$ and $β^2$.

8. (a) If $\dfrac{1}{2x+3} + \dfrac{1}{2x-3} = \dfrac{ax}{bx^2+c}$, where $x \ne \pm\dfrac{3}{2}$, find the value of a, b and c.
   (b) If the rate of interest of a borrower is increased from 3% to 4.5%, he will pay $3,600 more by simple interest in three years. How much money did he borrow?

9. Solve for x from the equation $\log_2(6.4x) = \log_5 625x$

10. Express the following in terms of trigonometric ratios of acute angles. Hence use mathematical tables to find their values.
    (a) $\sin 125°$
    (b) $\cos 225°$
    (c) $\tan 140°$

11. Find the values of x, y and z from the following circle.

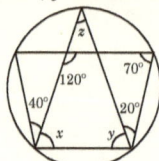

Figure 2.6.11.

12. Given the formula $A = P\left(1 + \dfrac{R}{100}\right)^n$.
    (a) Make n the subject of the formula.
    (b) Find n when $R = 3.5$ and $A = 1.5P$.

13. (a) The sum of the first ten terms of an arithmetic progression is 295 and the sixth term is four the sum of the first term and 1. Find the first term and the common difference.
    (b) The last term of a geometric progression is 2^{n+2} where n is the number of terms of the progression. Find the sum of the first ten terms of this progression.

14. The diameter of a spherical ball is nine times the diameter of a marble. What is the difference in terms of r if r is the radius of a marble between their
    (a) volumes?
    (b) areas?

15. Given that y varies directly as $(x+2)$, and that $y = 10$ when $x = 2$.
    (a) Find the equation which express y in terms of x,
    (b) From the formula, complete the table below.

x	4	—	8
y	—	20	—

    (c) Draw the graph of y against x on the first quadrant.

16. Find the area of the shaded region from the figure 2.6.16 below if O is the centre of the circle ABC, AB = 15 cm and BC = 8 cm.

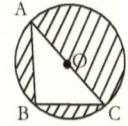

Figure 2.6.16.

17. Given the following triangle ABC.

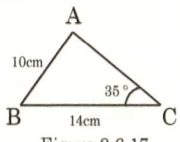

Figure 2.6.17.

  (a) Solve this triangle.
  (b) Find the area of the triangle.

18. The distance from a person of height 1.5 m to the building decreases by 10 m when the angle of elevation from the top of his head changes from 30° to 60°. Calculate
  (a) the distance from the building to a person when the angle of elevation from the top of his head was 30°.
  (b) the height of the building.

19. (a) Find the distance between points A(35°E, 20°N) and B(35°E, 25°S) on the earth's surface.
  (b) What is the distance on the latitude 40°N between 41°E and 15°W?

20. (a) The scale of the map is 1:50,000. Calculate
    (i) the actual distance in kilometres if the distance on the map is 5.8 cm, and
    (ii) the area of the region on the land if it is 6.5 cm² on the map.
  (b) If $\log_2 b = p$ and $\log_3 b = q$, find $\log_b 1.125$ in terms of p and q.

Section B

21. PQRST is a right pyramid in which PQRS is a square base of side 10 cm, and the side ST = 12 cm.

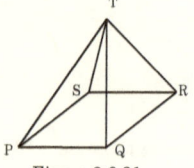

Figure 2.6.21.

Calculate
(a) the height of the pyramid,
(b) the angle that TQ makes with the base,
(c) the angle that the plane TPS makes with the base, and
(d) the surface area of the pyramid.

22. (a) From the information given below, prepare the trial balance as on 31st May, 2007.

Name of account	Amount in shs
Cash	252,000
Capital	$30x$
Purchases	265,000
Sales	288,650
Furniture and Fittings	x
Discount received	1,780
Discount allowed	$0.25x$
Insurance	3,550
Wages	$0.2x$
Creditors	3,580
Debtors	2,560
Opening stock	$5x+6,400$
Closing stock	$5x-6,000$

(b) Use the trial balance that you have drawn in (a) above to determine
   (i) the cost of goods available for sale,
   (ii) the cost of sales,
   (iii) the gross profit, and
   (iv) the net profit.

23. A person has 2.5 kg of iron and 1.65 kg of aluminium for making spoons and knives. In order to make a spoon, he needs 50 g of iron and 45 g of aluminium. To make a knife, he needs 60 g of iron and 30 g of aluminium. If the profit of each spoon and knife are shs.750 and shs.600 respectively,
   (a) How many number of each item should he make in order to maximize his profit?
   (b) What is his maximum profit?

24. (a) The polygon whose area is 8 cm² has been transformed by a transformation T_1, to the polygon whose area is 40 cm². If the matrix of T_1 is $\begin{pmatrix} x-1 & 2 \\ 2x & 3+x \end{pmatrix}$, find the value of x.
   (b) A linear transformation T_2, takes any point $P(x,y)$ onto $P'(x',y')$. Given that $x' = 2x+y$ and $y' = -x+2y$. Find:-
      (i) The matrix of transformation T_2.
      (ii) The image of the point (3,-3) when transformed by T_2.

25. (a) The shaded region on the figure 2.6.25 below represents a relation R. Study the figure carefully, and then answer the following questions.

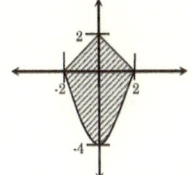

Figure 2.6.25.

   (i) From the figure, state the domain and range of R.
   (ii) Find the relation which expresses R.
   (b) Given that $\sin 10° = x$ and $\cos 10° = y$, show that

   $$\tan 55° = \frac{y+x}{y-x}$$

Test 7

Section A

1. Boyle's law states that *"The volume of a gas at a fixed mass is inversely proportional to its pressure at a constant temperature"*. If 10 cm³ of a gas supports 500 Pascals of a pressure, calculate
   (a) the pressure of the gas when the volume is 6.5 cm³.
   (b) the volume of the gas when the pressure is 250 Pascals.

2. In a certain school of forty-two students, each student must study at least one of the subject between physics, biology and chemistry. Nineteen students study physics, twenty-three study biology and twenty-five study chemistry. Seven students study all the three subjects, ten study both physics and biology and thirteen study both chemistry and biology.
   (a) Represent the information in a Venn diagram.
   (b) How many students study both physics and chemistry?
   (c) How many students study only one subject?

3. (a) Without using any mathematical aid, evaluate
   (i) $8^{\log_2 3}$ (ii) $6^{\log_{216} 8}$
   (b) A group of ten men can eat 5 kg of rice in 2 hours. How much rice can be eaten in 4 hours if four men leave the group?

4. (a) Factorize completely $x^3 - 7x + 6$.
   (b) Use the result obtained in (a) above to solve the equation $x^3 - 7x + 6 = 0$.

5. Given that $x^2 = 3\cos^2\theta$, $y = \sin\theta$ and $x+y = 2$. Find the value of x and y, then find θ if $0° \leq \theta \leq 90°$.

6. (a) The points (-1,-7), (2x,8) and (x,2) are collinear. Find the value of x.
   (b) Find the equation of a line which passes through the points in (a) above.

7. (a) Given that $f(x) = \dfrac{2x-5}{x^2 - 3x - 10}$. Find the domain of f.
   (b) A person has three ropes of lengths 20 m, 25 m and 30 m. He wants to cut those ropes exactly into pieces of equal lengths without any remains.
   (i) What is the longest possible lengths of the pieces?
   (ii) How many pieces can he get in (b)(i)?

8. From the figure 2.7.8 below, AC is a diameter of the circle. AE is a tangent at A and it is 12 cm long, BX = 3 cm. BD extended to E such that DE = 8 cm. Calculate

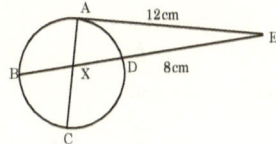

Figure 2.7.8.

(a) the length of the chord BD,
(b) the radius of the circle,
(c) the area of the circle.

9. Illustrate the relation R = {(x,y): y ≥ 2x+1 and y < 4} graphically, and hence state the domain and range of R.

10. In the figure 2.7.10 below, the radii of the circles whose centres are A and B, are 8 cm and 6 cm. The length PQ = 10 cm. Calculate the size of the shaded area common to the two circles.

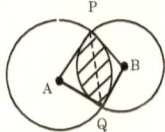

Figure 2.7.10.

11. Ali gains an average of shs.75,000 for 12 hours and 30 minutes for his job.
    (a) What is his rate of gain?
    (b) How much does he earn for 45 hours of his job?
    (c) How long should he work so as to gain shs.50,000?

12. An aircraft whose airspeed is 300 km/h has to fly in the direction of 089°, but the wind is blowing from N45°W at a speed of 120 km/h. Calculate
    (a) the ground speed of an aircraft,
    (b) the drift, and
    (c) the track.

13. On the figure 2.7.13 below, L, M and N are the midpoints of lines $\overline{PQ}, \overline{PR}$ and $\overline{QR}$ respectively.
    (a) Find in terms of **a** and **b** the vectors
       (i) $\overrightarrow{QR}$ (ii) $\overrightarrow{PN}$ (iii) $\overrightarrow{QM}$ (iv) $\overrightarrow{LN}$
    (b) What is the relation between vectors $\overrightarrow{LN}$ and $\overrightarrow{PR}$?

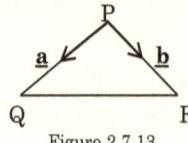

Figure 2.7.13.

14. (a) Find the sum to infinity for the series

    $$8.06 + 0.806 + 0.0006 + 0.00006 + \ldots$$

    (b) The arithmetic and geometric means of two numbers are 35 and 21 respectively. Find the numbers.

15. Given that $Q = 5P^n R$.
    (a) Make n the subject of the formula.
    (b) Find the value of n if $Q = 5.08$, $R = \dfrac{151}{523}$ and $P = 45.8$ and write the answer into three significant figures.

16. (a) The regular polygon has an interior angle of 135°. Find
       (i) the exterior angle of the polygon, and
       (ii) the number of sides of the polygon. State its name.
    (b) If the polygon in (a) above is inscribed in a circle of radius 10 cm, calculate
       (i) the area of the polygon, and
       (ii) the perimeter of the polygon.

17. (a) Calculate the speed of the earth's rotation about its axis if it completes the rotation in 24 hours.
    (b) Find the shortest distance on the earth's surface between point A(20°N, 38°E) and B(15°S, 112°W).

18. From the triangle below, find the possible angles and sides which are missing.

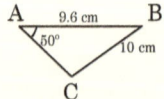

Figure 2.7.18.

19. Solve for θ such that $0° \leq \theta \leq 360°$ from the following equations:
    (a) $2\tan^2\theta = 5\tan\theta - 2$ (b) $12\sin\theta + 4\cos^2\theta = 9$
20. Find a in terms of b the following equations:
    (a) $\log_b a + 4 = 2\log_a b$
    (b) $\log_a b + \log_b a = 3b$ if $(\log_b a)^2 = 2b\log_b a$

Section B

21. From the rectangular prism below, the length AB = 10 cm, AR = 15 cm and PR = 12 cm. Calculate
    (a) the length CR,
    (b) the length AD,
    (c) the angle between plane ABRS and plane SRCD, and
    (d) the volume and surface area of the prism.

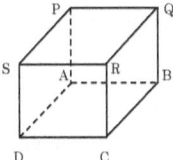

Figure 2.7.21.

22. The scores in geography examination obtained by 30 students in a certain class were recorded as follows:-

    48 63 56 38 39 69 63 41 59 70
    65 71 45 36 56 23 47 36 13 76
    36 44 38 45 48 18 56 39 66 52

    (a) Prepare a frequency distribution table with class intervals 11–20, 21–30, etc.
    (b) Calculate the mean and median mark.
    (c) Find the modal class and modal mark.

23. (a) The function f is defined as $f(x) = x - [x]$
        (i) Draw the graph of f.
        (ii) State the domain and range.
    (b) Determine the values of the following:-
        (i) $[-5]$
        (ii) $[2.8]$
        (iii) $[-\pi]$

24. (a) Find the inverse of the matrix $\begin{pmatrix} 2 & 1 \\ 1 & -3 \end{pmatrix}$.

(b) Use the answer obtained in (a) above to solve the system of equations
$$\begin{cases} 2x + y = 4 \\ x - 3y = 9 \end{cases}$$

(c) Point (x,y) is first reflected along the line $x+y = 0$ followed by the translation $\begin{pmatrix} 3 \\ -5 \end{pmatrix}$ and gives the final image at (a,b).

   (i) Write the equation in matrix form for the information.

   (ii) Find the image of point (3,-4) by using the equation obtained in (a) above.

25. (a) Two ordinary dice are rolled together. Find the probability of turning up two numbers whose sum is
   (i) exactly equal to 8.
   (ii) less than 6.
   (iii) greater than 7.

(b) A number in the form of $\dfrac{a}{b}$ where $b \neq 0$ is formed by using a denominator from the set of digits {1,3,5,7} and numerator from a set of digits {2,4,6,8,10}. Find the probability of forming a proper fraction or a whole number.

Tests for form 3 and 4

Test 8

Section A

1. (a) The area A of a sphere is proportional to the square of its radius, r. What the ratio must the radius being altered if the area increased by 44%?
   (b) a is partly constant and partly varies inversely as b. The product and the sum of the values of a when $b = 3$ and $b = 5$ are 35 and 12 respectively.
   (i) Write the equation connecting a and b.
   (ii) Find a when $b = 4$.
   (iii) Find b when $a = 5$.

2. Calculate the area of the following trapezium.

   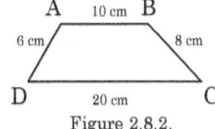

   Figure 2.8.2.

3. The price of a radio which includes a VAT is shs.51,750. If the rate of VAT is 25%, find the price of the radio before the VAT was added.

4. In a village of hundred people, the number of people who eat meat is eighty, who eat rice is eighty-one and those who eat beans are eighty-two. The number of people who eat rice and meat is sixty-eight, who eat meat and beans is sixty-seven and those who eat beans or rice but not meat is twenty. The number of people who eat all three kinds of food is fifty-nine. Find the number of people who eat
   (i) other foods than those mentioned above, and
   (ii) meat or beans but not rice.

5. The figure 2.8.5 below is a sphere in which part of it has been removed away leaving the cross sectional area of 300 cm². If point X is the maximum perpendicular height from the cross sectional area, which is 12 cm high, calculate
   (a) the diameter of the sphere,
   (b) the volume of the whole sphere before the part was removed away.

   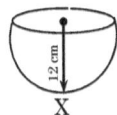

   Figure 2.8.5.

6. A ship is traveling on a course of 300° in still water at a speed of 50 km/h. There is a current of water of 15 km/h in the direction of 045°. Calculate the actual speed, drift and track.

7. (a) Find the area of a regular polygon of twenty sides inscribed in a circle of radius 15 cm.
   (b) Calculate the area of a triangle whose sides are 8 cm, 15 cm and 21 cm.

8. The ship leaves port P(30°N, 25°E) and sails due east for 20 hours to point Q(30°N, 60°E). Calculate
   (a) the distance that the ship traveled from P to Q, and
   (b) the speed of the ship.

9. The first, third and fifth terms of a series S is obtained by adding the corresponding terms of an arithmetic progression and a geometric progression of common ratio 3 are 19, 69 and 439. Find
   (a) the common difference of an arithmetic progression and the first terms of both arithmetic and geometric progressions,
   (b) the eighth term of S, and
   (c) the nth term of S.

10. (a) Find the equation of a perpendicular bisector of the line segment AB if point A is (3,7) and point B is (-2,9).
    (b) Find the equation of a line which passes through the point (3,5), which is
      (i) parallel;
      (ii) perpendicular;
      to the line whose equation is $5x+3y = 4$.

11. (a) The point (5,-4) is reflected to (-4,-4). Find the equation of the line of reflection.
    (b) The translations T_1 and T_2 have vectors $\begin{pmatrix} 2 \\ 5 \end{pmatrix}$ and $\begin{pmatrix} 3 \\ -6 \end{pmatrix}$ respectively. T_1 takes the point A(2,1) to the point B and T_2 takes the point B to the point C.
      (i) What are the coordinates of C?
      (ii) Find the translation that takes point C to A.
      (iii) State whether T_1T_2 is commutative or not.

12. Show that (a) $\dfrac{1}{\log_a x} + \dfrac{1}{\log_b x} = \dfrac{1}{\log_{ab} x}$

    (b) $\log_a 0.001 = -\dfrac{3}{\log_{10} a}$

13. The polynomial function $f(x) = ax^3 + bx^2 - 4x - c$ has the same remainders when divided by $(x+1)$ and $(x-2)$. Show that $3a+b = 4$, hence find the value of c if $f(x)$ leave the remainder equals to 15 when divided by $(x-3)$.

14. Given that $\dfrac{2a^2 + 5ab}{b^2} = 3$, find the value of $\left(\dfrac{a}{b}\right)^2$.

15. (a) Prove that the sum of interior angles of n-sided regular polygon is given by $180°(n-2)$. Hence find the sum of interior angles of a regular heptagon.
    (b) Find the number of sides of a regular polygon whose interior angle is $156°$.

16. Find the numerical value of the expression below:

$$\dfrac{5(2^{n+4}) + 4^{n+3} \times \dfrac{1}{2^n}}{3(2^{n+5}) - 6(2^{n+2})}$$

17. (a) Given the equation $v = \dfrac{1}{\sqrt{t^2 + 4t - 1}}$, make t the subject of the formula, then find t if $v = 0.5$.
    (b) Rationalize the denominator $\dfrac{\sqrt{3} - \sqrt{2}}{2 - \sqrt{6}}$, hence evaluate it if $\sqrt{2} = 1.414$.

18. The relation $R = \{(x,y): y = x^2 - 3\}$.
    (a) Find the domain and range of R.
    (b) Sketch the graph of R.
    (c) Find R^{-1}.

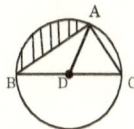

Figure 2.8.18.

19. (a) On the circle shown on figure 2.8.18 above, BC is a straight line and $AD = CD = BD$. Use this relationship to prove that $\angle BAC = 90°$.
    (b) Find the area of shaded region from the figure in (a) above if the length of an arc AC = 20 cm and the length of line BC = 30 cm.

20. (a) If an operation * over real numbers a and b is defined as a*b = 2a+b². Find
    (i) 2*(3*1) (ii) (x*y)*z
    (b) By completing the square, solve the quadratic equation
    $$\frac{x+1}{2x-3} = \frac{2x+2}{3x+2}$$

Section B

21. Find the volume of the frustum below, if O is the centre of the small circle and B is the centre of the large circle.

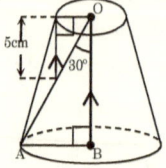

Figure 2.8.21.

Given OA = 12 cm, ∠AOB = 30°.

22. (a) (i) How many terms are there in the series below?
$$\begin{pmatrix} 2 & 1 \\ -1 & 3 \end{pmatrix} + \begin{pmatrix} 4 & 2 \\ -2 & 6 \end{pmatrix} + \begin{pmatrix} 8 & 4 \\ -4 & 12 \end{pmatrix} + \ldots + \begin{pmatrix} 128 & 64 \\ -64 & 192 \end{pmatrix}$$

    (ii) Find the sum of the series in a(i) above.

    (b) Point P(2,-3) is transformed by the matrix $\begin{pmatrix} 2 & 0 \\ -3 & 1 \end{pmatrix}$ followed by the reflection along the y-axis. Find the image of point P after this double transformation.

23. (a) A box contains eight red balls, ten blue balls and seven green balls. If a ball is picked at random from the box, find
    (i) the probability of picking a blue ball.
    (ii) the probability of picking a red or green ball.
    (iii) the number of blue balls to be added in the box to make the probability of picking a blue ball equal to $\frac{4}{9}$
    (iv) the number of blue balls to be removed in a box to make the probability of picking a red ball equal to 0.4.
    (b) X and Y are the independent events such that P(X) = 0.3 and P(X∪Y) = 0.65. Find P(Y).

24. (a) (i) Plot the graph of $f(x) = 2^x$.
    (ii) From the graph plotted in (a) above, state the domain and range.
    (iii) Find the inverse of $f(x)$.
    (iv) Is f one-to-one function?

    (b) A function f is defined as $f(x) = \dfrac{1}{2-x}$. Find the domain and range of f.

25. (a) Calculate the resultant speed from the figure 2.8.25(a) below.

    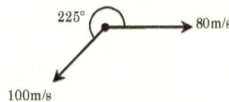

    Figure 2.8.25(a).

    (b) Juma, Faki and Jecha pulled an object, each one in the direction as shown below. Juma and Jecha pulled the object by the forces of 60 Newton and 70 Newton respectively. Calculate the force that Faki used to pull the object if the magnitude of resultant force between them is 38 Newton.

    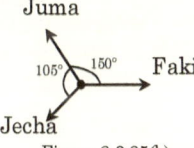

    Figure 2.8.25(b).

Test 9

Section A

1. (a) Show that $a^{\log_a b} = b$
   (b) If $\log 500 = 2.6990$, find $\log \dfrac{1}{5}$

2. Find b on the figure 2.9.2

   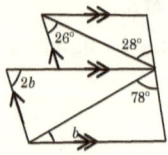

   Figure 2.9.2.

3. The terms of a certain series are arranged and grouped in the following way.

   $$2+(4+8)+(16+32+64)+(128+256+512+1024)+\ldots$$

   Find the first term of the kth group and the sum of the terms in the kth group.

4. (a) A box of biscuits costs shs.6,000. If a sales tax of 125% is imposed, what will be the price of a box of biscuits?
   (b) A shopkeeper sells a suit for shs.50,000 and makes a profit of 25%. Calculate the buying price of the suit.

5. A pilot wishes to direct the aircraft to North direction. The air is blowing to East and its speed is 50 km/h, and the speed of the aircraft is 120 km/h. Calculate
   (a) the resultant speed of the aircraft, and
   (b) the direction that a pilot should set in order to travel directly into North.

6. A relation R is defined as $R = \{(x,y): y = 2^{x-1}\}$.
   (a) Find the domain and range of R.
   (b) Find R^{-1}
   (c) Determine $R^{-1}\left(\dfrac{1}{8}\right)$

7. Newton's second law of motion states that *"The force F Newton acting on a body is proportional to the acceleration a m/s it produces"*. If a force of 33 Newton produces an acceleration of 0.75 m/s²,
   (a) Find F in terms of a.
   (b) What force would accelerate the body at 2.5 m/s²?
   (c) What acceleration is produced by a force of 66 Newton?
   (d) Draw the graph of F against a.

8. (a) Calculate the length of a chord of a circle of radius 17 cm which is 8 cm rom the centre of the circle.
   (b) A regular pentagon is inscribed in a circle and the area of shaded part in the pentagon is 47.56 cm² as shown on the figure 2.9.8 below. Find the total area of shaded region from the figure.

Figure 2.9.8.

9. (a) Calculate the area and perimeter of a square whose diagonal is 25 cm long.
   (b) The population of a certain country of 1.55×10^7 people increases by 15% each year. How many people were in the country seven years ago?
   (Write the answer into three significant figures)

10. (a) By using the value of $\tan 15° = 2 - \sqrt{3}$, find the value of
    (i) $\tan 75°$ (ii) $\tan 105°$

    (b) Given that $\cos\theta = 0.2$, without using tables, evaluate $\dfrac{\sin\theta\cos\theta}{\tan\theta}$.

11. The lines $2y = 3x + 12$ and $2x + y + 1 = 0$ intersect at point X. The line $\dfrac{x}{a} + \dfrac{y}{b} = 1$ where a and b are x- and y-intercepts, passes through the point X and it is perpendicular to the line $2y = 4x - 1$. Find the values of a and b.

12. (a) The second term of an infinite and convergent geometric series whose sum is 27 is equal to the third term of an infinite and convergent geometric series whose sum is 36. If the first terms of these series are the same, find the common ratio of each series.
    (b) Find the total interest obtained when £25,000 deposited in a bank at 2.5% after two and a half a year interest compounded semiannually.

13. A and B are two points on the earth's surface. Point A is at (35°S, 57°E) and point B is at (35°S, 123°W).
    (a) Calculate the distance between A and B along
       (i) the great circle.
       (ii) the parallel of latitudes.
    (b) Find the difference in length between the answer obtained in a(i) and that of a(ii) above.

14. How many numbers of three digits can be formed from the digits 1,2,3,4,5 and 6,
    (a) If the digits can be repeated in the same number?
    (b) If the digits are not repeated in the same number?
    (c) Will be less than 400 in (b) above?
    (d) Will be even numbers in (c) above?

15. A businessman from Kenya visited Tanzania. From the bureau de change, he exchanged kes.50,000 for tzs.850,000.
    (a) What is the exchange rate in:-
       (i) tzs per kes?
       (ii) kes per tzs?
    (b) How much kes can be obtained from tzs.8320?
    (c) How much tzs can be obtained from kes.25,000?

16. (a) The polynomial function $f(x) = x^3 - px^2 + 3x + q$ leaves the remainders 14 and -10 when divided by $(x - 3)$ and $(x+1)$ respectively. Find the values of p and q.
    (b) If $f(x) = x+1$ and $g(x) = x^2 - 1$, find
       (i) $f \circ g$
       (ii) $\dfrac{g(x)}{f(x)}$ and solve for x if $\dfrac{g(x)}{f(x)} = 0$.

17. (a) Find x given that $12x = 6.34 - \dfrac{1}{8}\log 0.6$ correct to two significant figures.
    (b) Evaluate $\log\left(\dfrac{18.56}{12.64} + \dfrac{11.24}{\sqrt{15.6}}\right)$

18. (a) Factorize completely the following expressions:
       (i) $x^2 - y^2 - 4x + 4y$ (ii) $2a^2 + b + 2a^2 b + 1$
    (b) Simplify; $\left(\dfrac{(2x^2 - 32)(4x + 8)}{x^2 + 5x + 4}\right) \div (x^2 - 2x - 8)$

19. (a) From the points A and B which are 60 m apart, and in the same plane with a tower, the angles of elevation of the top of the tower are 45° and 55° respectively. Find the height of the tower.
    (b) Find the angles in triangle whose sides are 18 cm, 20 cm and 24 cm long.

20. (a) Solve for θ from the equation $2\sin^2\theta = 3(1 - \cos\theta)$ if $0° \leq \theta \leq 360°$.
   (b) Draw the graph of $f(x) = \sin x$, and hence state the range.

Section B

21. ABCD is a tetrahedron. Given that AB = 6 cm, BC = AC = 5 cm and AD = BD = CD = 8 cm. Calculate
   (a) the height DO of the tetrahedron,
   (b) the angle CD makes with the base ABC, and
   (c) the angle between the planes BAD and ABC.

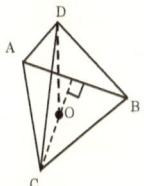

Figure 2.9.21.

22. Record the following transactions in the ledger.

2003, Jan 1	Commenced business with cash	$15,000
2	Purchased goods for cash	$7,000
5	Sold all goods for cash	$13,000
9	Purchased goods for cash	$6,500
13	Paid carriage on goods sold	$150
18	Sold goods for cash	$9,000
21	Paid wages	$250

   Balance and close the accounts.

23. A bookseller needs to put physics books of dimensions 20 cm by 14 cm by 2 cm each and mathematics books of dimensions 18 cm by 15 cm by 2.5 cm each in a box of dimensions 60 cm by 64 cm by 10 cm. Each physics book weighs $\frac{1}{3}$ kg and each mathematics book weighs $\frac{3}{8}$ kg. He has to put the box on his shoulders and take it to his shop, but his maximum ability to hold it is 22.5 kg by mass. He would sell each physics book and each mathematics book at shs.6,000 and shs.7,000 respectively. The box weighs 500g when empty. If he asks for the advice from you how would he put the books in the box in order to get the maximum amount of money for the books that he can put in the box, how would you advise him?

24. (a) Three events A, B and C are such that P(A) = 0.3, P(C) = 0.5, P(A∪B) = 0.58 and P(B∪C) = 0.9. If A and B are independent events, show that B and C are mutually exclusive events.

(b) Given that $f(x) = \begin{cases} |x| & if\ |x| < 2 \\ x^2 & if\ x \geq 2 \\ x+2 & if\ x \leq -2 \end{cases}$

Sketch the graph of f and state the domain and range.

25. (a) If $\underline{u} = 3i+2j$ and $\underline{v} = -2i+j$, find the matrix of a linear transformation if $T(2\underline{u}) = i - j$ and $T(3\underline{v}) = j - i$.

(b) Find the equation of the image of the line $2x+3y = 8$ after translated by the translation $T = \begin{cases} x' = x+2 \\ y' = y-3 \end{cases}$

(c) What is the image of the point A(3,6) when rotated by 60° clockwise direction about the origin?

Tests for form 3 and 4

Test 10

Section A

1. (a) If the radius of the base of a cylinder is increased by 10%, what percentage must the height decreases by if the volume remains constant?
   (b) a varies directly as the square of b and inversely as the square root of c. If a is increased by 10% and b decreased by 15%, find the percentage change in c.

2. (a) A tank can be filled by pipe P alone in 4 hours, pipe Q in 5 hours and pipe R in 6 hours. How long would it take if the pipes works together to fill the tank?
   (b) One of the roots of the cubic equation $x^3 - 2x^2 + tx + 6 = 0$ is -2. Find the value of t, and hence find the other roots.

3. Find the shaded area of the following figure, if X and Y are the centres of the circles which are 10 cm apart.

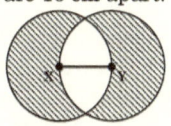

Figure 2.10.3.

4. A boat traveling at 10 km/h in still water is steered in the direction N48°E. It is carried off course by a current of 8 km/h in the direction 120°. Calculate the boats
   (a) resultant speed.
   (b) direction.

5. From the figure 2.10.5 below, PQRS is a parallelogram such that RS is the diameter of the circle PSTR. TB is a tangent at T, $\angle PTS = 35°$ and $\angle TBS = 40°$. Find
   (a) $\angle SPQ$ (c) $\angle SPT$
   (b) $\angle PQR$ (d) $\angle PST$

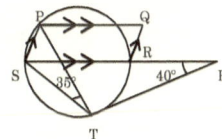

Figure 2.10.5.

6. The relation R is defined as R = {(x,y): 2y ≤ x+2, y ≥ 1 and x ≤ 3}.
   (a) Draw the graph which expresses the relation R and shade the required region.
   (b) From the graph, state the domain and range.

7. Given that x and y are the acute angles, and that $\sin x = \frac{\sqrt{2}}{2}$ and $\cos y = \frac{\sqrt{3}}{2}$, find the value of the following:
   (a) $\frac{\cos(x+y)}{\sin(x+y)}$
   (b) $\tan(x-y)$

8. Calculate the shortest distance from the line $y = \frac{2}{5}x + 7$ to the line $y = \frac{2}{5}x - 10$.

9. The first, second and fourth non-zero terms of an arithmetic progression forms the second, third and fourth terms of a geometric progression respectively. If the first term of the arithmetic progression is 6, find
   (a) the common difference of the arithmetic progression.
   (b) the common ratio and the first term of the geometric progression.
   (c) the sum of the first ten terms of both arithmetic and geometric progressions.

10. (a) In the first year, Ali earned $10,000 from an investment. The amount he earned increased at the annual rate of 3%. How much money to the nearest unit altogether will Ali receive from his investment in the first ten years?
    (b) Given that a, $2b$ and c are the consecutive terms of a geometric progression while a, $\frac{5}{2}b$ and c are the consecutive terms of an arithmetic progression. Find the common ratio of the geometric progression.

11. Use the substitution for $(x + \frac{1}{x}) = y$ to solve for x from the equation $4x^4 + 2x^3 - 22x^2 + 2x + 4 = 0$

12. A certain ship leaves the port P(75°N, 20°W) at 7:35 a.m. and sails due North to the port Q along the great circle. If the length of the journey of the ship is 7779 km, find
    (a) the position of port Q.
    (b) the length of the journey in nm.
    (c) the time taken for the ship to reach Q if the speed of the ship is 100 knot.
    (d) when the ship reach port Q.

13. The size of an interior angle of a regular polygon is three times the size of its exterior angle.
    (a) Find (i) the exterior angle.
              (ii) the number of sides of the polygon.
    (b) If this polygon is inscribed in a circle of area 20 cm², calculate (i) the length of each side of the polygon.
              (ii) the area and perimeter of the polygon.

14. The ratio of the perimeter and area of a rhombus is 5:12 cm and the ratio of its diagonals is 3:4. Find
    (a) the lengths of the diagonals of the rhombus.
    (b) the area of the square whose perimeter is the same as that of rhombus, and hence find the lengths of diagonals of that square.

15. (a) Without using tables, find the value of the following:
    (i) $\cos 15°$ (ii) $\sin\dfrac{11\pi^c}{12}$ (iii) $\tan\left(\dfrac{5\pi^c}{9} - \dfrac{17\pi^c}{36}\right)$

    (b) Use the results obtained in (a) above to find the value of
    $$\dfrac{\sin\tfrac{11}{12}\pi^c \times \cos 15°}{\tan\left(\tfrac{5}{9}\pi^c - \tfrac{17}{36}\pi^c\right)}$$

16. From the figure 2.10.18 below, O is the centre of the circle ABC, CT is a tangent at C, and OD is perpendicular to the chord AB. The secant ABT meets with tangent CT at T. Prove that $\overline{AB}^2 = 4(\overline{OC}^2 - \overline{OD}^2)$. Then if OC = 5 cm, OD = 3 cm and AT = 12 cm, calculate:

    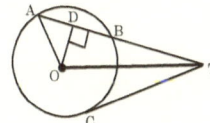

    Figure 2.10.18.

    (a) AD
    (b) TB

17. Mr. Salim bought a car for shs 8,000,000. After a while, he decided to sell it and reduces its cost. He gave the car to Kombo so as to sell it for him. When Kombo sold it, he increased the selling price that Mr. Salim decided to sell it by 16%, therefore there is only 27.5% loss left from the buying price of the car. Find
    (a) the price that Mr. Salim decided to sell the car, and
    (b) the percentage loss of the price that Mr. Salim decided to sell the car.

18. (a) AB is a straight line of length 12 cm. The point X divides AB internally in the ratio 2:1 and Y divides AB externally in the ratio 3:1. Find the lengths AX, BX, AY and BY.
    (b) On the figure 2.10.20 below, ABCD is a square and DEC is an equilateral triangle. What is the name of triangle CBE?

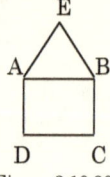

Figure 2.10.20.

Then, calculate ∠BCE.

19. (a) If $a^{\frac{1}{3}} = \sqrt{b} \times c^2$ and $a^{\frac{1}{2}} \cdot \sqrt[3]{b} = c^2$, find b in terms of c only.
    (b) Given that $a = x^3 y$ and $b = \sqrt[3]{y} \times x^2$. Express x in terms of a and b.
    (c) If $2^x = 3^2$, find the value of x.

20. (a) Solve the inequality $x^2 - 2x \leq 3$.
    (b) The area and perimeter of the semicircle are given by the formulae $A = \frac{1}{2}\pi r^2$ and $P = \pi r + 2\pi$ respectively. Find P if A = 18.25 units square and $\pi = 3.14$.

Tests for form 3 and 4

Section B

21. When Dr. Paul commenced business on 1ˢᵗ March, 2008, he had the stock of £20,000 and the capital £300,000.

    March 2: purchased goods for cash £100,000,
          3: sold goods for cash £120,000,
          6: purchased goods for cash £80,000,
         10: returned unsaleable goods that he bought to the owner £5,000,
         15: sold goods for cash £160,000,
         18: bad goods that he sold returned back to him £8,000,
         19: drawing in cash £50,000,
         20: purchased goods for cash £70,000,
         26: sold goods for cash and paid wages £110,000 and £8,000/= respectively.

    Stock in hand on 31ˢᵗ March £25,000.
    (a) Enter the above transactions in the cash account.
    (b) Prepare the trading, and the profit and loss accounts.
    (c) Construct the balance sheet as on 31ˢᵗ March, 2008.

22. (a) The probability that James will pass the exam is twice the probability that Asha will fail the exam, and the probability that Asha will pass the exam is three times the probability that Seif will fail the exam. If the probability that either Asha or Seif will pass the exam is $\frac{59}{64}$, find the probability that
    (i) both Asha and Seif will pass the exam.
    (ii) all Asha, Seif and James will fail the exam.
    (iii) either James or Seif will pass the exam.
    (iv) either James or Asha will fail the exam.

    (b) Given that $f(x) = -3x^2 + 4x - 5$.
    (i) Write f in the form of $f(x) = a(x+b)^2 + c$ where a, b and c are real numbers.
    Determine
    (ii) the turning point
    (iii) the axis of symmetry
    (iv) the maximum or minimum value
    (v) the range of f.

Tests for form 3 and 4

23. (a) The ship P leaves the point X(30°S, 20°W) at 8:30 a.m. and sails due North, and the ship Q leaves the point Y(45°N, 20°W) two hours later and sails due South. If the speed of the ships P and Q are 80 knots and 100 knots respectively,
    (i) when will the ships meet?
    (ii) at what point will they meet?
    (b) The height and radius of a cone are 12 cm and 8 cm respectively. Calculate
    (i) its volume, and
    (ii) its surface area.

24. (a) (i) Find the inverse of the matrix $\begin{pmatrix} 4 & 2 \\ 1 & 3 \end{pmatrix}$.

    (ii) Use the answer in a(i) above to solve the following simultaneous equations:
    $$\begin{cases} 4x + 2y = 3 \\ x + 3y = -2 \end{cases}$$

    (b) Find the matrix X if matrix $A = \begin{pmatrix} 1 & 2 \\ -1 & 4 \end{pmatrix}$ and $B = \begin{pmatrix} 10 & -1 \\ 14 & -5 \end{pmatrix}$ from the equation $AX = B$.

    (c) The line $2x+y+4 = 0$ is first reflected along the line $x+y = 0$ followed by the translation with vector $\begin{pmatrix} -2 \\ 4 \end{pmatrix}$. Find the image of the line in the form of $\frac{x}{a} + \frac{y}{b} = 1$ where a and b are real numbers, and then state the x- and y- intercepts.

25. The figure 2.10.25 below is a right frustum of a square base. The height of the frustum is 10 cm. Calculate
    (a) the angle between the planes ABQP and SPQR,
    (b) the angle between the plane DCQP and the base, and
    (c) the volume of the frustum.

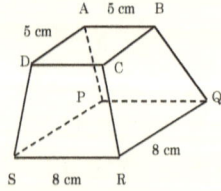

Figure 2.10.25.

Tests for form 3 and 4

Solutions for Tests 1 And 2

Test 1

1. Given $n = 10$, $r = 10$ cm

   (a) Area $= \dfrac{1}{2}r^2 n \sin\left(\dfrac{360°}{n}\right)$

   $= \dfrac{1}{2} \times 10^2 \times 10 \sin\left(\dfrac{360°}{10}\right)$

   $= 500\sin 36°$

   $= 293.9$ cm^2

   (b) Perimeter $= 2rn\sin\left(\dfrac{180°}{n}\right)$

   $= 2 \times 10 \times 10 \sin\left(\dfrac{180°}{10}\right)$

   $= 200\sin 18° = 61.80$ cm

2. (a) $\underline{c} = \underline{a}+\underline{b} = 3i - 2j + 3i + 7j$

   $= 6i + 5j$

   (b) $2\underline{a} - \underline{b} = 2(3i - 2j) - (3i + 7j)$

   $= 3i - 11j$

   $|2\underline{a} - \underline{b}| = \sqrt{3^2 + 11^2} = \sqrt{130}$

3. $0.35555... - 0.23232... = 0.12323...$

   Now, let $x = 0.1\dot{2}\dot{3} \Rightarrow 100x = 12.3\dot{2}\dot{3}$

   $99x = 12.2 \Rightarrow x = \dfrac{61}{495}$

4.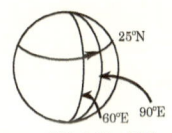

   $\theta = 90° - 60° = 30°$.

   $l = \dfrac{\theta \pi R \cos\alpha}{180°} = \dfrac{30° \times 3.14 \times 6370 \cos 25°}{180°} = 3021$ km

   Speed $= \dfrac{3021}{24} = 125.9$ km/h

5. $a^2 \alpha \dfrac{b}{\sqrt{c}} \Rightarrow a^2 = \dfrac{kb}{\sqrt{c}} \Rightarrow k = \dfrac{a^2 \times \sqrt{c}}{b} = \dfrac{2^2 \times \sqrt{4}}{3} = \dfrac{8}{3}$

   Now, $b = \dfrac{a^2 \times \sqrt{c}}{k} = \dfrac{4^2 \times \sqrt{9} \times 3}{8} = 18$

6. $G_1 = 5$, $G_n = 1215$ and $S_n = 1820$.

   $S_n = \dfrac{G_1(r^n - 1)}{r - 1} \Rightarrow 1820 = \dfrac{5(r^n - 1)}{r - 1} \Rightarrow 1820r - 1820 = 5r^n - 5$

   $\Rightarrow 1820r - 5r^n = 1815$

   But $G_n = G_1 r^{n-1}$

   $1215 = 5r^{n-1}$, then $r^n = 243r$

   Then, $1820r - 5(243r) = 1815 \Rightarrow 605r = 1815 \Rightarrow r = 3$

7. (a) a ≈ 22.4 and b ≈ 44.8

   (b) $\dfrac{a \times b}{a^2} = \dfrac{22.4 \times 44.8}{22.4^2} = \dfrac{22.4^2 \times 2}{22.4^2} = 2$

7. (a) $27 = \dfrac{1}{2}n(n-3)$

   $\Rightarrow 54 = n^2 - 3n$
   Solve for n and get $n = 9$ or $n = -6$
   The number of diagonals must be 9.
   ∴ The polygon is Nonagon.

   (b) $5 = \dfrac{1}{2}n(n-3) \Rightarrow 10 = n^2 - 3n$

   Solve for n and get $n = 5$ or $n = -2$
   The number of diagonals must be 5.
   ∴ The polygon is Pentagon.

8. $\begin{pmatrix} 2 & 2 \\ 2 & -3 \end{pmatrix} \begin{pmatrix} x \\ y \end{pmatrix} = \begin{pmatrix} 10 \\ 5 \end{pmatrix}$

   The determinant of $\begin{pmatrix} 2 & 2 \\ 2 & -3 \end{pmatrix} = (-3 \times 2) - (2 \times 2) = -10$

   Then $-\dfrac{1}{10}\begin{pmatrix} -3 & -2 \\ -2 & 2 \end{pmatrix}\begin{pmatrix} 2 & 2 \\ 2 & -3 \end{pmatrix}\begin{pmatrix} x \\ y \end{pmatrix} = -\dfrac{1}{10}\begin{pmatrix} -3 & -2 \\ -2 & 2 \end{pmatrix}\begin{pmatrix} 10 \\ 5 \end{pmatrix}$

   $\begin{pmatrix} x \\ y \end{pmatrix} = \begin{pmatrix} 0.3 \times 10 + 0.2 \times 5 \\ 0.2 \times 10 - 0.2 \times 5 \end{pmatrix} = \begin{pmatrix} 4 \\ 1 \end{pmatrix}$

   ∴ $x = 4$ and $y = 1$

9. The expression must be a perfect square (has equal roots)

   By using $b^2 = 4ac$ where a = 4, b = (p − 25) and c = 25

   $(p - 25)^2 = 4 \times 4 \times 25 \Rightarrow p^2 - 50p + 625 = 400 \Rightarrow p = 45$ or $p = 5$

10. (a) $\sin^4 a - \cos^4 a = (\sin^2 a)^2 - (\cos^2 a)^2$
   $= (\sin^2 a + \cos^2 a)(\sin^2 a - \cos^2 a)$

   But $\sin^2 a + \cos^2 a = 1$

   ∴ $\sin^4 a - \cos^4 a = (\sin a + \cos a)(\sin a - \cos a)$

   (b) $\sin^2 A + \cos^2 A = 1$

   $\Rightarrow 0.6^2 + \cos^2 A = 1 \Rightarrow \cos A = 0.8$

   (i) $\sin B = \sin(45° - A) = \sin 45° \cos A - \cos 45° \sin A$

   $= \dfrac{\sqrt{2}}{2}(0.8) - \dfrac{\sqrt{2}}{2}(0.6) = \dfrac{\sqrt{2}}{10}$

   (ii) $\cos B = \cos(45° - A) = \cos 45° \cos A + \sin 45° \sin A$

   $= \dfrac{\sqrt{2}}{2}(0.6) + \dfrac{\sqrt{2}}{2}(0.8) = \dfrac{7\sqrt{2}}{10}$

   (iii) $\tan B = \dfrac{\sin B}{\cos B} = \dfrac{\sqrt{2}}{10} \div \dfrac{7\sqrt{2}}{10} = \dfrac{1}{7}$

Tests for form 3 and 4

11. $2a+9b = 4$ ...(i) $\Rightarrow a = \dfrac{4-9b}{2}$

    $\dfrac{6b-c}{7} = \dfrac{1}{4}$ ...(ii)

    $24b - 4c = 7 \Rightarrow c = \dfrac{24b-7}{4}$

    But $4a+8c = 4$...(iii)

    Then $4\left(\dfrac{4-9b}{2}\right) + 8\left(\dfrac{24b-7}{4}\right) = 4 \Rightarrow 8 - 18b + 48b - 14 = 4$

    $\Rightarrow 30b = 10 \Rightarrow b = \dfrac{1}{3}$

    $a = \dfrac{4-9b}{2} = \dfrac{4-9\left(\frac{1}{3}\right)}{2} = \dfrac{1}{2}$ and $c = \dfrac{24b-7}{4} = \dfrac{24\left(\frac{1}{3}\right)-7}{4} = \dfrac{1}{4}$

---

12. $\log a^3 b = 3.5 \Rightarrow 3\log a + \log b = 3.5$...(i)

    $\log a^2 \cdot \sqrt{b} = 1.5 \Rightarrow 2(\log a + \dfrac{1}{2}\log b) = 1.5 \times 2$

    $\Rightarrow 4\log a + \log b = 3$...(ii)

    Take equation (ii) – (i) and get $\log a = \bar{1}.5$
    But $\log b = 3 - 4\log a$...(ii)

    $\log b = 3 - 4(\bar{1}.5) = 5$

    (a) $\log (ab)^3 = 3\log ab = 3(\log a + \log b) = 3(\bar{1}.5 + 5) = 13.5$

    (b) $\log \sqrt{\dfrac{b}{a}} = \dfrac{1}{2}(\log b - \log a) = \dfrac{1}{2}(5 - \bar{1}.5) = 2.75$

---

13. If a^{-2}, b^{-2} and c^{-2} are in G.P, then

    $\dfrac{b^{-2}}{a^{-2}} = \dfrac{c^{-2}}{b^{-2}} \Rightarrow \dfrac{a^2}{b^2} = \dfrac{b^2}{c^2} \Rightarrow b^2 = ac$

    This is a geometric mean, therefore a, b and c are also in GP.

    Again, from $b^2 = ac$

    $\Rightarrow 2\log b = \log a + \log c \Rightarrow \log b = \dfrac{\log a + \log c}{2}$

    This is an arithmetic mean, therefore $\log a$, $\log b$ and $\log c$ are the terms in AP.

14. (a) $\dfrac{\log_7 15}{\log_3 8} = \dfrac{\log 15}{\log 7} \div \dfrac{\log 8}{\log 3} = \dfrac{\log 15}{\log 7} \times \dfrac{\log 3}{\log 8} = \dfrac{1.1761 \times 0.4771}{0.8451 \times 0.9031}$

Number	log
1.1761	0.0704
0.4771	$\bar{1}.6786$
Numerator	$\bar{1}.7490$
0.8451	$\bar{1}.9269$
0.9031	$\bar{1}.9557$
Denominator	$\bar{1}.8826$
7.352×10^{-1}	$\bar{1}.8664$

$\therefore \dfrac{\log_7 15}{\log_3 8} = 0.7352$

(b) Let
$$x = \dfrac{\sqrt[3]{0.0122} \times 10.543}{(11.2)^2 \times 0.00123}$$

$\log x = \dfrac{1}{3}\log 0.0122 + \log 10.543 - 2\log 11.2 - \log 0.00123$

$\quad = \bar{1}.3621 + 1.0228 - 2.0984 - \bar{3}.0899 = 1.1966$

$\Rightarrow x =$ antilog $1.1966 = 15.73$

---

15. $\begin{pmatrix} a & 2b-3 \\ 5 & c-b \end{pmatrix} = \begin{pmatrix} 4 & 9 \\ 5 & 3 \end{pmatrix}$

$a = 4,\ 2b - 3 = 9 \Rightarrow b = 6$
$c - b = 3 \Rightarrow c - 6 = 3 \Rightarrow c = 9$

---

16. (a) $\dfrac{1}{2}, \dfrac{3}{5}, \dfrac{6}{11}, \dfrac{12}{23} \ldots$

The next terms are
$\dfrac{23+1}{23+24} = \dfrac{24}{47},\ \dfrac{47+1}{47+48} = \dfrac{48}{95}$

(b) $T_1 = \dfrac{13}{9} = \dfrac{13-0}{9+4\times 0}$

$T_2 = \dfrac{13-1}{9+4\times 1},\ T_3 = \dfrac{11}{17} = \dfrac{13-2}{9+4\times 2},\ T_4 = \dfrac{10}{21} = \dfrac{13-3}{9+4\times 3}$

Then, $T_n = \dfrac{13-(n-1)}{9+4(n-1)} = \dfrac{14-n}{5+4n}$

---

17. The points are (0,2) and (-2,0).

(i) Gradient, $m = \dfrac{2-0}{0+2} = 1$

(ii) From $y = m(x - x_1) + y_1 \Rightarrow y = 1(x+2) + 0 \Rightarrow y = x+2$

(iii) $m_1 m_2 = -1$ if $m_1 = 1$, then $m_2 = -1$
From $y = m(x - x_1) + y_1 \Rightarrow y = -1(x - 3) + 3 \Rightarrow y = -x + 6$

18. (a) $n(X \cup Y) + n(X \cup Y)' = n(\mu)$
$n(X \cup Y) = 17 - 5 = 12$
(b) $n(X \cap Y) = n(X) + n(Y) - n(X \cup Y)$
$= 7 + 8 - 12 = 3$
(c) $n(X') + n(X) = n(\mu)$
$n(X') + 7 = 17 \Rightarrow n(X') = 10$

19. $n(S) = 8$
(a) $P(E) = \dfrac{n(E)}{n(S)} = \dfrac{5}{8}$
(b) $P(E) = \dfrac{n(E)}{n(S)} = \dfrac{6}{8} = \dfrac{3}{4}$

20. $f(x) = -2x^2 + 4x - 3 = -2(x^2 - 2x) - 3 = -2(x^2 - 2x + 1^2) - 3 + 2$
$\Rightarrow f(x) = -2(x - 1)^2 - 1$
(a) (1,-1)   (b) Maximum value, $y = -1$
(c) $x = 1$ (d) Range = $\{y: y \leq -1\}$

21. $10y + 40° + 9y + 15° + 11y + 20° + 45° + 270° + 5y + 10° + 15y = 900°$ (Hexagon)
$50y + 400° = 900°$
$50y = 500°$
$y = 10°$

22. (a)

Class limit	Tally mark	Frequency	Class mark	d=x–A	fd	Cum. freq.
20–30	///	3	25	-35	-105	3
30–40	//// /	6	35	-25	-150	9
40–50	//// //	7	45	-15	-105	16
50–60	//// ////	10	55	-05	-50	26
60–70	//// //	7	65	05	35	33
70–80	////	5	75	15	75	38
80–90	//	2	85	25	50	40
		$\Sigma f = 40$			250	

(b)

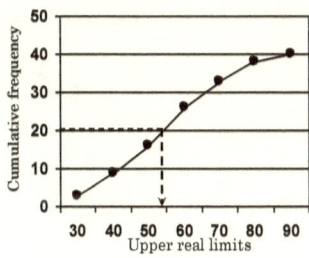

Median = 54

(c) Mean = $A + \dfrac{\Sigma fd}{\Sigma f} = 60 - \dfrac{250}{40} = 53.75$

23. (a) $(x+2)(3x+5) - 7(4x-2) = 0$
$3x^2+5x+6x+10 - 28x+14 = 0$

$3x^2 - 17x+24 = 0 \Rightarrow x = 3$ or $x = \dfrac{8}{3}$

(b) (i) $AB = \begin{pmatrix} 2 & 3 \\ 4 & 8 \end{pmatrix}\begin{pmatrix} 0 & -1 \\ 2 & -3 \end{pmatrix} = \begin{pmatrix} 0+6 & -2-9 \\ 0+16 & -4-24 \end{pmatrix} = \begin{pmatrix} 6 & -11 \\ 16 & -28 \end{pmatrix}$

(ii) $\Delta A = 2\times 8 - 3\times 4 = 16 - 12 = 4$

$A^{-1} = \dfrac{1}{4}\begin{pmatrix} 8 & -3 \\ -4 & 2 \end{pmatrix} = \begin{pmatrix} 2 & -\tfrac{3}{4} \\ -1 & \tfrac{1}{2} \end{pmatrix}$

(c) $2(A^{-1}+A) - B = 2\left[\begin{pmatrix} 2 & -0.75 \\ -1 & 0.5 \end{pmatrix} + \begin{pmatrix} 2 & 3 \\ 4 & 8 \end{pmatrix}\right] - \begin{pmatrix} 0 & -1 \\ 2 & -3 \end{pmatrix}$

$= 2\begin{pmatrix} 4 & 2.25 \\ 3 & 8.5 \end{pmatrix} - \begin{pmatrix} 0 & -1 \\ 2 & -3 \end{pmatrix} = \begin{pmatrix} 8 & 5.5 \\ 4 & 20 \end{pmatrix}$

---

24. (a) Dr   Cash A/c   Cr

Date	Particulars	Folio	Amount ($)	Date	Particulars	folio	Amount ($)
Jan 1	Capital		260,000	Jan 2	Purchases		70,000
3	Sales		90,000	7	Purchases		30,000
27	Sales		60,000	13	Pack. mat.		5,000
				19	Trn. char.		3,500
				21	Purchases		20,000
				24	Insurance		6,000
				31	balance	c/d	275,500
			410,000				410,000
Feb 1	Balance	b/d	275,500				

Dr   Sales A/c   Cr

Date	Particulars	Folio	Amount ($)	Date	Particulars	folio	Amount ($)
Jan 31	Balance	c/d	150,000	Jan 3	Cash		90,000
				27	Cash		60,000
			150,000				150,000
				Feb 1	Balance	b/d	150,000

(b)   TRIAL BALANCE

Accounts	Dr ($)	Cr ($)
Cash	275,500	-
Sales	-	150,000
Purchases	120,000	-
Capital	-	260,000
Packing material	5,000	-
Transport charges	3,500	-
Insurance	6,000	-
	410,000	410,000

Test 2

1. (a) Range = $\{y: -1 \leq x \leq 1\}$
   (b) $(\sin x - \cos x)^2 + (\sin x + \cos x)^2$.
   $= \sin^2 x - 2\sin x \cos x + \cos^2 x + \sin^2 x + 2\sin x \cos x + \cos^2 x$
   $= 2\sin^2 x + 2\cos^2 x = 2(\sin^2 x + \cos^2 x)$
   $= 2 \times 1 = 2$

2. $A_5 = a + 4d = 31 \ldots(i)$
   where a is the first term and d is the common difference.
   $S_{12} = \dfrac{12}{2}(2a + 11d) \Rightarrow 400 = 6(2a + 11d)$
   $\Rightarrow 400 = 12a + 66d \ldots(ii)$
   Solve for equations (i) and (ii) and get $a = \dfrac{223}{9}$ and $d = \dfrac{14}{9}$

3. (a)

   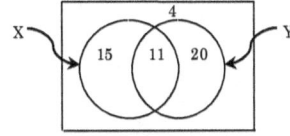

   (b) (i) $15 + 11 = 26$
   (ii) $11 + 20 = 31$
   (iii) 11
   (iv) $11 + 20 + 4 = 35$

4. Let the numbers be a and b.
   Then, $10 = \dfrac{a+b}{2} = a + b = 20 \ldots(i)$
   $8 = \sqrt{ab} = 64 \ldots(ii)$
   $b = 20 - a \ldots(i)$
   $ab = 64 = a(20 - a) \Rightarrow 20a - a^2 = 64 \Rightarrow a = 16$ or $a = 4$
   Then, if $a = 16$, $b = 4$ and if $a = 4$, $b = 16$
   $\therefore$ The numbers are 4 and 16.

5. Let $x = \dfrac{(25.32)^2 \times \sqrt{45.38}}{(5.834)^3 \times \sqrt[3]{84.12}}$

   $\log x = 2\log 25.32 + \dfrac{1}{2}\log 45.38 - (3\log 5.834 + \dfrac{1}{3}\log 84.12)$

   $= 2 \times 1.4035 + \dfrac{1}{2} \times 1.6569 - (3 \times 0.7660 + \dfrac{1}{3} \times 1.9249)$

   $= 2.8070 + 0.8285 - (2.2980 + 0.6416)$
   $= 3.6355 - 2.9396$
   $= 0.6959$
   Then $x = $ antilog $0.6959 = 4.965 \approx 4.97$ (2 decimal places)

Tests for form 3 and 4

6. Area $= \frac{1}{2}r^2 n \sin\left(\frac{360°}{n}\right) \Rightarrow 1175.6 = \frac{1}{2}r^2 \times 6\sin\left(\frac{360°}{6}\right)$

   $1175.6 = 3r^2 \sin 60° \Rightarrow r = 21.27$ cm

   Area of the circle $= \pi r^2 = 3.14(21.27)^2 = 1420.6$ cm².

   Circumference $= 2\pi r = 2 \times 3.14 \times 21.27 = 133.6$ cm

---

7. $f(x) = \dfrac{5}{x-a} + \dfrac{6+a}{x+2a} \Rightarrow f(2) = \dfrac{5}{2-a} + \dfrac{6+a}{2+2a} = \dfrac{1}{0}$

   $\dfrac{5(2+2a) + (2-a)(6+a)}{(2-a)(2+2a)} = \dfrac{1}{0} \Rightarrow (2-a)(2+2a) = 0$

   $a = 2$ or $a = -1$

   Then, $f(x) = \dfrac{5}{x-2} + \dfrac{8}{x+4}$ or $f(x) = \dfrac{5}{x+1} + \dfrac{5}{x-2}$

   $\therefore f(4) = \dfrac{5}{4-2} + \dfrac{8}{4+4} = 3\dfrac{1}{2}$ or $f(4) = \dfrac{5}{4+1} + \dfrac{5}{4-2} = 3\dfrac{1}{2}$

---

8. (a) (i) AB $= \sqrt{(0.7+0.8)^2 + (-1+0.2)^2} = \sqrt{2.25+0.64} = \sqrt{2.89} = 1.7$

   (ii) AC $= \sqrt{(0.7+0.2)^2 + (-1-0.2)^2} = \sqrt{0.81+1.44} = \sqrt{2.25} = 1.5$

   (b) (i) $\left(\dfrac{-0.8-0.2}{2}, \dfrac{-0.2+0.2}{2}\right) = (-0.5, 0)$

   (ii) $\left(\dfrac{0.7-0.2}{2}, \dfrac{-1+0.2}{2}\right) = (0.25, -0.4)$

   (c) (i) Gradient, $m = \dfrac{0.2+0.2}{-0.2+0.8} = \dfrac{0.4}{0.6} = \dfrac{2}{3}$

   From $y = m(x - x_1) + y_1$

   $y = \dfrac{2}{3}(x + 0.2) + 0.2$

   $y = \dfrac{2}{3}x + \dfrac{1}{3} \Rightarrow 3y = 2x + 1 \Rightarrow -2x + 3y = 1$

   $\therefore \dfrac{x}{\left(-\frac{1}{2}\right)} + \dfrac{y}{\left(\frac{1}{3}\right)} = 1$

   (ii) Use the same procedures as in c(i) above and get the equation

   $\dfrac{x}{\left(-\frac{1}{20}\right)} + \dfrac{y}{\left(-\frac{1}{15}\right)} = 1$

---

9. Let x litres of juice contain 25% of sugar mixed with y litres of juice contain 35% of sugar in order to produce $(x+y)$ litres of juice which contain 28% of sugar.

   Then, $0.25x + 0.35y = 0.28(x+y)$
   $0.25x + 0.35y = 0.28x + 0.28y$
   $0.03x = 0.07y$
   $\dfrac{x}{y} = \dfrac{7}{3}$

   ∴ The two types of juice containing 25% and 35% of sugar will be mixed in the ratio 7:3 respectively.

Tests for form 3 and 4

10. (a) $\underline{t} = 2\underline{r} - \underline{s} = 2(2i - 4j) - (4i+j) = -9j$

    Then, $|\underline{t}| = \sqrt{(-9)^2} = 9$

    (b) $2x - 3y + 5 = 0 \Rightarrow y = \dfrac{2}{3}x + \dfrac{5}{3}$, then $m_1 = \dfrac{2}{3}$

    From $m_1 m_2 = -1$ (the lines are perpendicular)

    $\dfrac{2}{3} m_2 = -1 \Rightarrow m_2 = -\dfrac{3}{2}$

    From $y = m(x - x_1) + y_1$

    $y = -\dfrac{3}{2}(x+1) + 2 \Rightarrow 3x + 2y - 1 = 0$

    ∴ The equation is $3x + 2y - 1 = 0$

11.

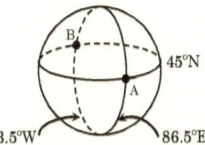

(a) Difference in degrees = $86.5° + 93.5° = 180°$

   Difference in time = 12 hours (half rotation)

   Then, the time at B = 2:28 p.m. − 12 hours = 2:28 a.m.

(b) (i) $\theta = 180° - (45° \times 2) = 180° - 90° = 90°$

   $l = \dfrac{\theta \pi R}{180°} = \dfrac{90 \times 3.14 \times 6370}{180} = 10{,}000$ km

   (ii) $\theta = 180°$, $\alpha = 45°$

   $l = \dfrac{\theta \pi R \cos \alpha}{180°} = \dfrac{180 \times 3.14 \times 6370 \cos 45°}{180} = 14{,}140$ km

---

12.
```
    kg       g       mg
    15      150      150
  ×            23.5
    356     028      525
    325.5   3525     3525
```
3525 mg = 3 g and <u>525 mg</u>

3525 g = 3 kg and 525 g, then 525 g + 3 g = 528 g

352.5 kg + 3 kg = 355.5 kg = 355 kg and 500 g

then, 528 g + 500 g = 1028 g = 1 kg and <u>28 g</u>

then, 355 kg + 1 kg = <u>356 kg.</u>

Tests for form 3 and 4

13. (a) Let the numbers be $2x$, $3x$, $4x$ and $5x$.
    Then, $2x+3x+4x+5x = 21 \Rightarrow x = 1.5$
    ∴ The numbers are $2(1.5) = 3$, $3(1.5) = 4.5$, $4(1.5) = 6$ and $5(1.5) = 7.5$

    (b) a:b:c:d = e:f:g:h
    Then, $\dfrac{a}{e} = \dfrac{b}{f} = \dfrac{c}{g} = \dfrac{d}{h} = k \Rightarrow a = ek, b = fk, c = gk$ and $d = hk$

    Then, $\dfrac{a+b+c+d}{e+f+g+h} = \dfrac{ek+fk+gk+hk}{e+f+g+h} = \dfrac{k(e+f+g+h)}{e+f+g+h} = k$

    Since $\dfrac{a}{e} = k$ and $\dfrac{a+b+c+d}{e+f+g+h} = k$

    Then, $\dfrac{a+b+c+d}{e+f+g+h} = \dfrac{a}{e}$ (hence proved)

---

14. If α and β are the roots, then the equation must be
    $(x - \alpha)(x - \beta) = 0$
    Then, $x^2 - (\alpha+\beta)x + \alpha\beta = 0$
    Again, from the equation $ax^2+bx+c = 0 \Rightarrow x^2 + \dfrac{b}{a}x + \dfrac{c}{a} = 0$

    Then, $-(\alpha+\beta) = \dfrac{b}{a} \Rightarrow \alpha+\beta = -\dfrac{b}{a}$ and $\alpha\beta = \dfrac{c}{a}$

    Then, $x^2 - (\alpha+\beta)x + \alpha\beta = x^2 + \dfrac{b}{a}x + \dfrac{c}{a}$

    $ax^2 - a(\alpha+\beta)x + a\alpha\beta = ax^2 + bx + c = 0$ (Hence shown)

---

15. (a) Let $x^2 = y$
    Then, $ay^2 + by + c = 0$

    From the quadratic formula, $y = \dfrac{-b \pm \sqrt{b^2 - 4ac}}{2a}$

    Then, $x^2 = \dfrac{-b \pm \sqrt{b^2 - 4ac}}{2a} \Rightarrow x = \sqrt{\dfrac{-b \pm \sqrt{b^2 - 4ac}}{2a}}$

    (b) From the formula above, when $a = 1$, $b = -13$ and $c = 36$

    $x = \sqrt{\dfrac{13 \pm \sqrt{(-13)^2 - 4 \times 1 \times 36}}{2}} = \sqrt{\dfrac{13 \pm \sqrt{25}}{2}} = \sqrt{\dfrac{13 \pm 5}{2}}$

    ∴ $x = \sqrt{\dfrac{13+5}{2}} = \pm 3$ or $x = \sqrt{\dfrac{13-5}{2}} = \pm 2$

---

16. $\angle BDC = \dfrac{1}{2}a$ (Angle at the circumference is half the angle at the centre)

    $\angle ADB + \angle BDC = 180°$ (Corresponding angles)

    $\angle ADB + \dfrac{1}{2}a = 180° \Rightarrow \angle ADB = 180° - \dfrac{1}{2}a$

    Also, $\angle ABD + \angle ADB + \angle BAD = 180°$ (the angles in triangle)

    $\angle ABD + 180° - \dfrac{1}{2}a + b = 180°$

    Then, $\angle ABD = \dfrac{1}{2}a - b$ (hence proved)

Tests for form 3 and 4

17. If the radius of the circle is r and the length of one side of a square is a, then,
    $\pi r^2 = a^2$ (Area of the circle = Area of the square)

    Then $a = r\sqrt{\pi}$

    The circumference of the circle = $2\pi r = 2 \times 3.14r = 6.28r$

    The perimeter of the square = $4a = 4 \times r\sqrt{\pi} = 4r\sqrt{3.14} = 7.09r$

    ∴ Circumference of the circle < Perimeter of the square (Shown)

---

18. (a) Let $A = \begin{pmatrix} 2 & 2 \\ 4 & 5 \end{pmatrix}$, then $|A| = 2 \times 5 - 2 \times 4 = 2$

    $A^{-1} = \frac{1}{2}\begin{pmatrix} 5 & -2 \\ -4 & 2 \end{pmatrix} = \begin{pmatrix} \frac{5}{2} & -1 \\ -2 & 1 \end{pmatrix}$

    (b) Let $B = \begin{pmatrix} 2 & -7 \\ -3 & 8 \end{pmatrix}$, then $|B| = 2 \times 8 - (-7 \times -3) = -5$

    $B^{-1} = -\frac{1}{5}\begin{pmatrix} 8 & 7 \\ 3 & 2 \end{pmatrix} = \begin{pmatrix} -\frac{8}{5} & -\frac{7}{5} \\ -\frac{3}{5} & -\frac{2}{5} \end{pmatrix}$

---

19. Let sides XY = $5x$ and YZ = $6x$

    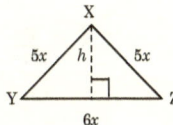

    By Pythagoras theorem, $(3x)^2 + h^2 = (5x)^2$
    $h^2 = 25x^2 - 9x^2 \Rightarrow h = 4x$

    Area = $\frac{bh}{2} \Rightarrow 108 = \frac{4x \times 6x}{2} \Rightarrow x = 3$ cm.

    Perimeter = $5x + 5x + 6x = 16x = 16 \times 3 = 48$ cm

---

20. $a \propto \frac{(b^2 + 2)}{\sqrt{c+1}} \Rightarrow a = k\frac{(b^2+2)}{\sqrt{c+1}} \Rightarrow k = \frac{a\sqrt{c+1}}{b^2+2} = \frac{3\sqrt{3+1}}{2^2+2} = 1$

    When $b = 4$ and $c = 2$, $a = \frac{(4^2+2)}{\sqrt{2+1}} = 6\sqrt{3}$

---

21. $(x-2)^2 + (x-1)^2 = x^2 \Rightarrow x = 5$ or $x = 1$

    But x must be equal to 5 only because in $x-2$ when $x = 1$ we will get -1.

    Surface area = Area of (ΔABC + ΔPQR + BCRQ + APRC + ABQP)

    Area of ΔABC = $\frac{1}{2}(x-2)(x-1) = \frac{1}{2} \times 3 \times 4 = 6$ cm²

    Area of ΔPQR = 6 cm²
    Area of BCRQ = $x \times 2x = 5 \times 10 = 50$ cm²
    Area of APRC = $(x-1)(2x) = 4 \times 10 = 40$ cm²
    Area of ABQP = $(x-2)(2x) = 3 \times 10 = 30$ cm²
    ∴ The total surface area = $(6 + 6 + 50 + 40 + 30)$ cm² = 132 cm².

22. (a)

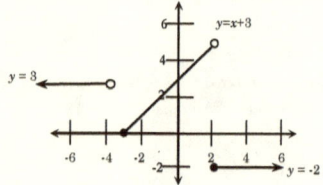

(b) Domain = {x: x < -4 or x ≥ -3}, Range = {y: y = -2 or 0 ≤ y ≤ 5}
(c) (i) f(3) = -2
   (ii) Doesn't exist.

23. (a)

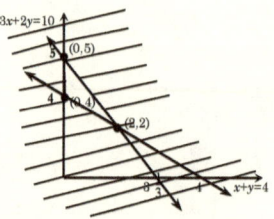

Corner points	F(x,y) = 4x+5y
(0,5)	4(0)+5(5) = 25
(2,2)	4(2)+5(2) = 18.
(0,4)	4(0)+5(4) = 20

For the minimum value, $x = 2$ and $y = 2$.

(b) Let x be the number of type A and y be the number of type B.

	Type A	Type B	Total
Cost (shs)	10,000	5,000	10,000
Number of gowns	x	y	14
Profit	4,500	2,500	

Constraints: $10000x+5000y \leq 100,000 \Rightarrow 2x+y \leq 20$,
$x+y \leq 14$, $x \geq 0$ and $y \geq 0$
Objective function is $F(x,y) = 4500x+2500y$

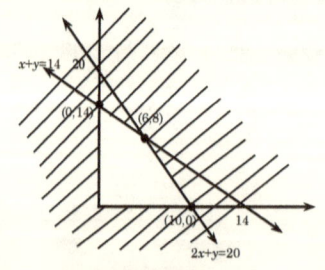

Corner points	F(x,y)=4500x+2500y
(0,14)	4500(0)+2500(14)=35,000
(6,8)	4500(6)+2500(8)=47,000
(10,0)	4500(10)+2500(0)=45,000

∴For the maximum profit, she could have bought 6 gowns of type A and 8 gowns of type B.

24. (a) (i) $\dfrac{4}{10} \times \dfrac{4}{10} = \dfrac{4}{25}$ (ii) $\dfrac{4}{10} \times \dfrac{6}{10} = \dfrac{6}{25}$

(b) Let A represents Ali and J represents Jane.
Then, (i) P(J)+P(J') = 1 where P(J') = 0.43
P(J)=1-0.43=0.57
Then, P(A∩J) = 0.65×0.57 = 0.3705
(ii) P(A)+P(A') = 1 ⇒ P(A') = 1 – 0.65 = 0.35
Then, P(A'∪J') = P(A')+P(J') – P(A'∩J')
But P(A'∩J') = 0.35×0.43 = 0.1505
∴P(A'∪J') = 0.35×0.43 – 0.1505 = 0.6295.

25.

Class intervals	Class-mark	Frequency	fx	Cum. frequency
1–20	10.5	18	189	18
21–40	30.5	33	100.5	51
41–60	50.5	36	1818	87
61–80	70.5	60	4230	147
81–100	90.5	45	4072.5	192
101–120	110.5	8	884	200
		$\sum f = 200$	$\sum fx = 12200$	

(a) Mean weight = $\dfrac{\Sigma fx}{\Sigma f} = \dfrac{12200}{200} = 61$ kg

(b) Median = $L + \left(\dfrac{\frac{N}{2} - \Sigma f_b}{f_m} \right) c$

Where L (Lower real limit of median class0 = 60.5

    c (Class size) = 20

    Σf_b (The sum of frequencies below the median class)

    f_m (Frequency of median class) = 60

    N (Total frequencies) = 200

Then, Median = $60.5 + \left(\dfrac{100 - 87}{60} \right) 20 = 64.83$ kg

(c) Mode = $L + \dfrac{t_1 c}{t_1 + t_2}$

Where, L (Lower real limit of modal class)

t_1 (The difference between the frequency of modal class and the frequency below it) = 60 – 36 = 24

t_2 (The difference between the frequency of modal class and the frequency above it) = 60 – 45 = 15

c (class size) = 20

Then, Mode = $60.5 + \dfrac{24 \times 20}{24 + 15} = 72.81$ kg

Tests for form 3 and 4

Answers from Test 3 – 10

Test 3

1. (a) $a = 4$ or $a = \dfrac{1}{9}$
   (b) $4x^2 - 20x + 25 = 0$
   or $x^2 + 30x + 225 = 0$
2. (a) 4.8 cm (b) 12.12 cm
   (c) 13.8 cm
4. (a) (2,-3)
   (b) $x = 1$ or $x = 2$ or $x = 3$
5. 132.5°W
6. (a) 17.32 m (b) 20 m
7. $a = 2$ and $b = 3$
8. (a) $\dfrac{ab + 9z - 12a}{b - 3}$
   (b) $\dfrac{a + z(b - 4)}{b - 3}$
   (c) $\dfrac{2a(b - 3) + (z - 4)(b - 4)}{2}$
9. 32.85 cm²
10. (a) $a = 2$, $b = 3$ and $r = 15$
    (b) $x = 1.53$
11. (a) $i + 9j$ (b) $\sqrt{82}$
12. $2x^2 - 9x + 36$
13. 20%
14. $\begin{pmatrix} -1 & 0 \\ 0 & 1 \end{pmatrix}$, (-3,5)
15. (a) $x = 0$ or $x = 0.63$
    (b) $2 < x \leq 6$
16. 75
17. $7,610
18. 5.47×10^{-1}
19. (a) 0.7002 (b) -0.6947
    (c) -0.9397
20. 36°

21. (i)

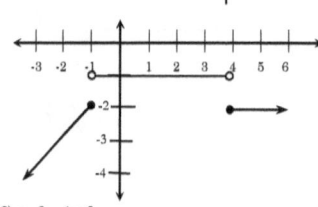

(ii) $f(-2) = -2 - 1 = -3$

(iii) Domain = $\{x: x \in \Re\}$, Range = $\{y: y = -1$ or $y \leq -2\}$

(iv) $f(x)$ is not one-to-one function because some of the elements of the range are corresponding with more than one element of the domain.

22. (a) 30.73 (b) 29.5 (c) 23.19
23. (a) 10 cm (b) $10\sqrt{2}$ cm (c) 45° (d) 59°05'
24. (a) (i) $\dfrac{1}{6}$ (ii) $\dfrac{5}{18}$ (iii) $\dfrac{13}{18}$ (b) 0.25
25. (a)

Dr   Cash A/c   Cr

Date	Particulars	Folio	Amount (shs)	Date	Particulars	Folio	Amount (shs)
Jan, 1	Capital		8,000,000	Jan, 2	Purchases		2,000,000
4	Sales		2,500,000	9	Purchases		1,500,000
10	Sales		1,800,000	15	Furniture		800,000
22	Sales		3,200,000	16	Mot. veh.		1,500,000
				18	Purchases		1,080,000
				20	Salaries		900,000
				31	balance	c/d	7,720,000
			15,500,000				15,500,000
Feb 1	balance	b/d	7,720,000				

Tests for form 3 and 4

(b)

Dr			Cr
Purchases	4,580,000	Sales	7,500,000
Less: Closing stock	780,000		
Cost of sales	3,800,000		
Gross profit c/d	3,700,000		
	7,500,000		7,500,000
Salaries	900,000	Gross profit b/d	3,700,000
Net profit	2,800,000		
	3,700,000		3,700,000

(c) BALANCE SHEET

Capital	8,000,000	Fixed Assets	
Add: Net profit	2,800,000	Motor vehicle	1,500,000
		Furniture	800,000
		Current Assets	
		Stock	780,000
		Cash	7,720,000
	10,800,000		10,800,000

Test 4

1. $x = \frac{1}{2}$ and $y = -\frac{1}{3}$ or $x = \frac{1}{10}$ and $y = \frac{1}{9}$

2. $11,550$ cm^2

4. $\frac{6}{7}$

5. $f(x) = -2\left(x - \frac{3}{4}\right)^2 + \frac{33}{8}$

   (a) $\left(\frac{3}{4}, \frac{33}{8}\right)$

   (b) $x = \frac{3}{4}$

   (c) Maximum value, $y = \frac{33}{8}$

   (d) $\{y: y \leq \frac{33}{8}\}$

6. $a = 4$ and $b = 6$

7. (a) $V = h\left[\dfrac{\pi h^2 \pm h\sqrt{\pi(\pi h^2 + 2A)} + A}{2}\right]$

   (b) 308 cm^2

8. 10 packets

9. $a = 1$ and $b = -4$, $(x+1)$

10. (a) tzs.57,684 (b) kes.869.56
    (c) ugx.197.95

11. (a) 33,750 (b) 15 terms

12. 60.22 cm^2

13. (a) $132°$ (b) $66°$ (c) $48°$

14. (a) $5x - y + 7 = 0$ (b) $-\frac{17}{3}$

15. (a) 1:3

16. (a) $11i + 9j$ (b) $\sqrt{202}$
    (c) $39°17'$ (d) $\dfrac{11}{\sqrt{202}}$ and $\dfrac{9}{\sqrt{202}}$

17. (a)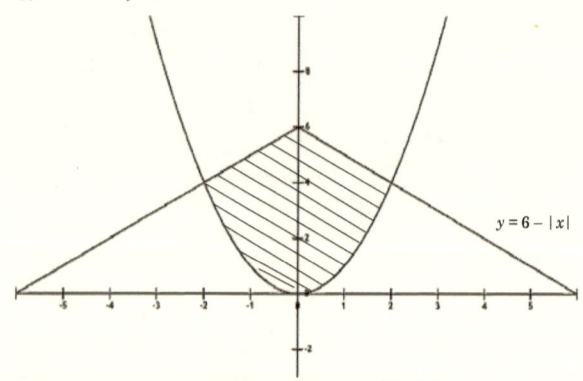

(b) Domain = {x: -2 ≤ x ≤ 2}, Range = {y: 0 ≤ y < 6}

Note: The shaded region represents the relation.

18. (a) $y = 2x - 1$ (b) $y = 2x - 9$
19. (a) $x = 5$ (b) 4cm (c) 34.5cm
20. (a) (i) {Natural numbers less than 20}
    (ii) {Multiples of 6 less than 20}, 16

    (b) (-3,-10)

21. 5 skirts and 6 blouses

22. (a)

Dr			Cash A/c				Cr
Date	Particulars	Folio	Amount ($)	Date	Particulars	folio	Amount ($)
Apr 1	Capital		20,000	Apr 1	Stock		2,000
3	Sales		10,000	3	Purchases		8,000
17	Sales		15,000	6	Wages		1,000
28	Sales		5,000	13	Purchases		10,000
				25	Salaries		4,000
				26	Insurance		1,500
				30	balance	c/d	23,500
			50,000				50,000
May 1	Balance	b/d	23,500				

(b) THE TRIAL BALANCE ACCOUNT

Account	Dr ($)	Cr ($)
Capital	-	20,000
Sales	-	30,000
Opening stock	2,000	-
Wages	1,000	-
Purchases	18,000	-
Salaries	4,000	-
Insurance	1,500	-
Cash	23,500	-
	50,000	50,000

Tests for form 3 and 4

(c)

Dr		Cr	
Opening stock	2,000	Sales	30,000
Purchases	18,000		
Cost of goods available	20,000		
Less: Closing stock	1,200		
Cost of sales	18,800		
Gross profit c/d	11,200		
	30,000		30,000
Wages	1,000	Gross profit b/d	11,200
Salaries	4,000		
Insurance	1,500		
Net profit	4,700		
	11,200		11,200

Balance sheet

Capital	20,000	Closing stock	1,200
Add: Net profit	4,700	Cash	23,500
	24,700		24,700

23. (a) 828.96 cm²   (b) 2076.6 cm³
24. (a) $x = 5$ (c) 48.5
25. (a)

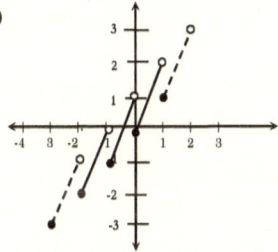

(b) Domain = {x: x is any real number},
Range = {y: y is any real number}

Test 5

1. (a) -1
   (b) $\theta = 60°$ or $90°$ or $270°$ or $300°$
2. 176.3 cm²
3. (a) 252.8 km/h   (b) 290°27′
   (c) 09°33′
4. (a) $x = 6$ or $x = -3$
   (b) (i) 0.2   (ii) 0.267
5. 20 cm and 14 cm
   If the triangle is an obtuse, the obtuse angle must be opposite to the longest side which is 20 cm. Then,
   $$\cos\theta = \frac{14^2 + 10^2 - 20^2}{2 \times 14 \times 10} = -0.3714$$
   $\Rightarrow \theta = 111°48′$
   ∴The triangle is an obtuse angled triangle (Hence shown)
6. (a) (i) 12.6   (ii) 12.570
   (b) 5.828
8. (a) 135   (b) 266
9. (a) -4,4 and 12   (b) 4,12 and 36
10. (a) $\frac{7}{9}$ (b) $\frac{233}{990}$ (c) $\frac{511}{495}$
11. (a) 174.4 km   (b) 126°34′
12. 55%
13. $x = 10$ and $y = 2$
14. (a) (ii) -5   (ii) 200   (iii) $x = \pm 5$
    (b) {x: -4 < x ≤ 8}
15. X = 4, 180, Y = 6,000 and Z = 60,000
16. (a) (i) 0.078   (ii) 0.0780   (b) 4
    (c) Thousandth
17. 1.8 hours
19. (a) $X = \begin{pmatrix} -16 & 36 \\ -24 & 18 \end{pmatrix}$
    (b) (i) $\begin{pmatrix} 0 & 0 \\ 0 & 0 \end{pmatrix}$ (ii) Zero matrix
20. (a) $a = 3$, $b = -5$
    (b) $p = 1$ and $q = 4$
    (c) $y = \frac{x}{3} + \frac{5}{3}$

Tests for form 3 and 4

21. (a) (i) 14 oranges and 12 mangoes
    (ii) 15 oranges and 12 mangoes
    (b) (i) tzs.2,840 (ii) tzs.2,940

22. (a) (i) (24,-2) (ii) (0,0)
    (b) $(2\sqrt{3}-1, \sqrt{3}+2)$
    (c) $Q = \begin{pmatrix} 1 & 3 \\ 2 & 1 \end{pmatrix}$

23. (a) (i) Domain = $\{x: x \in \Re\}$,
    Range = $\{y: y \geq 1\}$
    (ii) ± 2
    (b) (i)

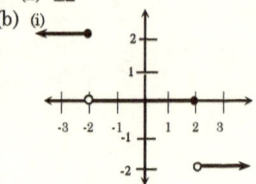

    (ii) $f(5) = -2$ and $f(-2) = 2$
    (iii) Domain = $\{x: x$ is any real number$\}$,
    Range = $\{-2, 0, 2\}$
    (iv) Is not, because some of the elements of the range corresponds with more than one element of the domain.

24. (a) $x = 5$ (b) 33.2

(c)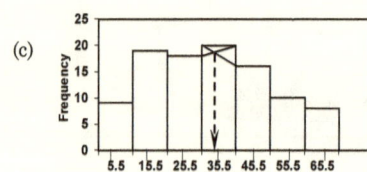

∴ Mode = 34

25. (a) (i) $\frac{1}{22}$ (ii) $\frac{19}{66}$ (iii) $\frac{5}{22}$

    (b) (i) 20 (ii) 12 (iii) 8 (iv) $\frac{3}{5}, \frac{2}{5}$

Test 6

1.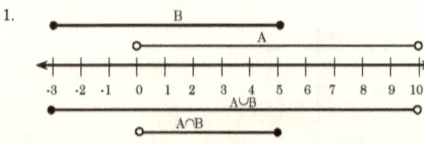

2. $\frac{\sqrt{6}-2}{3}$

3. (a) $a = 5$ (b) $x = -2.5$
4. (a) $x = 1.97$ (b) 8 notes of shs.500 and 2 of shs.2,000

5.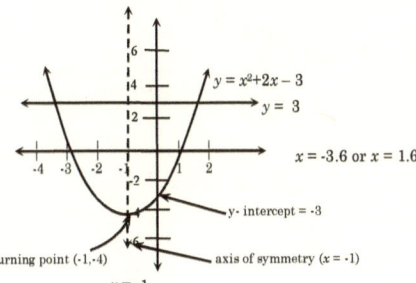

6. (a) 3:5, 1.91 cm
   (b) 11.45 cm^2, 31.75 cm^2
7. (a) ±4   (b) $16x^2+31x+49 = 0$
8. (a) a = 4, b = 4 and c = -9
   (b) $80,000
9. x = 5
10. (a) sin55°, 0.8192
    (b) -cos45°, -0.7071
    (c) -cos40°, -0.8391
11. x = 70°, y = 80° and z = 30°
12. (a) $n = \dfrac{\log A - \log P}{\log\left(1 + \dfrac{R}{100}\right)}$
    (b) n = 11.82
13. (a) 7 and 5 respectively
    (b) 8,184
14. (a) $\dfrac{2912\pi r^3}{3}$ (b) $320\pi r^2$

15. (a) $y = \dfrac{5}{2}x + 5$
    (b)

x	4	6	8
y	15	20	25

16. 166.87 cm^2
17. (a) A = 53°25′, B = 91°35′ and
        AC = 17.4 cm
        Or A = 126°35′, B = 18°25′ and
        AC = 5.51 cm
    (b) 70 cm^2 or 22.1 cm^2 respectively
18. (a) 15 m   (b) 10.16 m
19. (a) 5,000 km   (b) 4,767 km
20. (a) (i) 2.9 km   (ii) 1.625 km^2
    (b) $\dfrac{2p - 3q}{pq}$
21. (a) 9.7 cm   (b) 53°53′
    (c) 62°43′   (d) 318.2 cm^2

22. a)

NAME OF ACCOUNT	DR (shs)	CR (shs)
Cash	252,000	-
Capital	-	300,000
Purchases	265,000	-
Sales	-	288,650
Furniture and Fitting	10,000	-
Discount received	-	1,780
Discount allowed	2,500	-
Insurance	3,550	-
Wages	2,000	-
Creditors	-	3,580
Debtors	2,560	-
Opening stock	56,400	-
	594,010	594,010

(b) (i) shs.321,400   (ii) shs.277,400

(iii) shs.11,250   (iv) shs.4,980

23. (a) 20 spoons and 25 knives   (b) shs.30,000

24. (a) $x = 4$ or $x = -2$ (b) (i) $\begin{pmatrix} 2 & 1 \\ -1 & 2 \end{pmatrix}$ (ii) $(3,-9)$

25. (a) (i) Domain = $\{x: -2 \leq x \leq 2\}$ and Range = $\{y: -4 \leq y \leq 2\}$
    (ii) $R = \{(x,y): y \geq x^2 - 4 \text{ and } y \leq 2 - |x|\}$

Test 7

1. (a) 769.2 Pascals (b) 20 cm^3

2. (a)

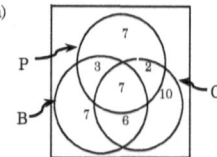

P represents Physics
C represents Chemistry
B represents Biology
(b) 9 students (c) 24 students

3. (a) (i) 27 (ii) 2 (b) 6 kg
4. (a) $(x+3)(x-2)(x-1)$
   (b) $x = 2$ or $x = 1$ or $x = -3$
5. $x = 1.5$ and $y = 0.5$ or $x = 4$ and $y = -2$, $\theta = 30°$
6. (a) $x = 2$ (b) $y = 3x - 4$
7. (a) Domain = $\{x: x \neq -2, x \neq 5\}$
   (b) (i) 5 m (ii) 15 pieces
8. (a) 10 cm (b) 5.67 cm
   (c) 101 cm^2

9.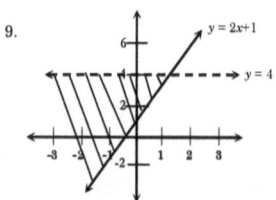

   Domain = $\{x: x \leq 1.5\}$,
   Range = $\{y: y \leq 4\}$
   Note: The shaded region represents the relation R

10. 30.8 cm^2
11. (a) shs.6000 per hour
    (b) shs.270,000
    (c) 8 hours and 20 minutes
12. (a) 393 km/h (b) 12°41'
    (c) 101°41'
13. (a) (i) $\mathbf{b} - \mathbf{a}$ (ii) $\frac{1}{2}(\mathbf{a}+\mathbf{b})$
    (iii) $\frac{1}{2}\mathbf{b} - \mathbf{a}$ (iv) $\frac{1}{2}\mathbf{b}$
    (b) $\overrightarrow{LN} = 2\overrightarrow{PR}$

14. (a) $\frac{133}{15}$ (b) 7 and 63
15. (a) $n = \frac{\log Q - \log 5R}{\log P}$
    (b) $n = 0.329$
16. (a) (i) 45° (ii) 8, Octagon
    (b) (i) 282.8 cm^2 (ii) 61.23 cm
17. (a) 1,667 km/h (b) 19,450 km
18. $\angle C = 47°20'$, $\angle B = 82°40'$ and $AC = 12.9$ cm
19. (a) $\theta = 26°34'$ or $63°26'$ or $206°34'$ or $243°26'$
    (b) $\theta = 30°$ or $150°$
20. (a) $a = b^{-2 \pm \sqrt{6}}$ (b) $a = \sqrt[6]{b}$
21. (a) 9 cm (b) 6.63 cm (c) 36°23'
    (d) 597 cm^3, 432.1 cm^2
22. (b) Mean = 48.2, Median = 47.6
    (c) 31–40 and 41–50, 40.5
23. (a) (i)

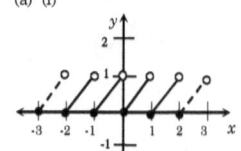

   (ii) Domain = $\{x: x$ is any real number$\}$
   Range = $\{y: 0 \leq y \leq 1\}$
   (b) (i) -5 (ii) 2 (iii) -4
24. (a) $-\frac{1}{7}\begin{pmatrix} -3 & -1 \\ -1 & 2 \end{pmatrix}$
    (b) $x = 3, y = -2$
    (c) (i) $\begin{pmatrix} a \\ b \end{pmatrix} = \begin{pmatrix} 0 & -1 \\ -1 & 0 \end{pmatrix}\begin{pmatrix} x \\ y \end{pmatrix} + \begin{pmatrix} 3 \\ -5 \end{pmatrix}$
    (ii) $(7,-8)$
25. (a) (i) $\frac{5}{36}$ (ii) $\frac{5}{18}$ (iii) $\frac{5}{12}$
    (b) $\frac{13}{20}$

Test 8
1. (a) $r_1:r_2 = 1:1.2$
   (b) (i) $a = 2 + \dfrac{15}{b}$ or $a = 10 - \dfrac{15}{b}$
   (ii) $\dfrac{23}{4}$ or $\dfrac{25}{4}$ respectively.
   (iii) 5 or 3 respectively.
2. $72\ cm^2$
3. shs.41,400
4. (i) No one (ii) 19
5. (a) 19.96 cm (b) 4162 cm^3
6. 48.3 km/h, $17°27'$ and $317°27'$ respectively
7. (a) 659.2 cm^2 (b) 46.43 cm^2
8. (a) 3368 km (b) 168.4 km/h
9. (a) 5, 14 and 5 respectively
   (b) 10,984 (c) $9 + 5(n + 3^{n-1})$
10. (a) $10x - 4y + 27 = 0$
    (b) (i) $5x + 3y - 30 = 0$
    (ii) $3x - 5y + 16 = 0$
11. (a) $x = \dfrac{1}{2}$
    (b) (i) (7,0) (ii) $\begin{pmatrix} -5 \\ 1 \end{pmatrix}$
    (iii) It is commutative
13. $a(-1)^3 + b(-1)^2 - 4(-1) - c = a(2)^3 + b(2)^2 - 4(2) - c$,
    Then get $9a + 3b = 12 \Rightarrow 3a + b = 4$
    and get $c = 9$
14. 0.25 or 9
15. (a) Heptagon has $900°$
    (b) 15
16. 2
17. (a) $t = \dfrac{-2v \pm \sqrt{5v^2 + 1}}{v}$,
    $t = 1$ or $t = -5$
    (b) $-\dfrac{\sqrt{2}}{2}$, -0.707

18. (a) Domain = $\{x: x \in \Re\}$,
    Range = $\{y: y \geq -3\}$
    (b)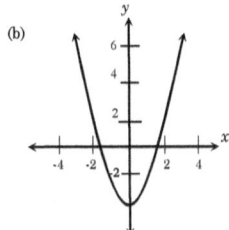
    (c) $R^{-1} = \{(x,y): y = \sqrt{x+3}\ \}$
19. (b) 93.9 cm^2
20. (a) (i) 53 (ii) $4x + 2y^2 + z^2$
    (b) $x = -1$ or $x = 8$
21. 670.6 cm^3
22. (a) (i) 7 (ii) $\begin{pmatrix} 254 & 127 \\ -127 & 381 \end{pmatrix}$
    (b) (-4,-9)
23. (a) (i) $\dfrac{2}{5}$ (ii) $\dfrac{3}{5}$ (iii) 2 (iv) 5
    (b) $\dfrac{1}{2}$
24. (a) (i)

   $f(x) = 2^x$

   (ii) Domain = $\{x: x \in \Re\}$,
   Range = $\{y: y > 0\}$
   (iii) $f^{-1}(x) = \log_2 x$ (iv) Yes.
   (b) Domain = $\{x: x \neq 2\}$,
   Range = $\{y: y \neq 0\}$
25. (a) 71.31 m/s
    (b) 75.5 N or 64.7 N

Test 9
1. (b) $\bar{1}.3010$
2. $b = 33.3°$
3. $\sqrt{2^{k^2 - k + 2}}\left(\sqrt{2^{k^2 + k + 1}} - \sqrt{2^{k^2 - k + 2}}\right)$
4. (a) shs.7,500 (b) shs.40,000

Tests for form 3 and 4

5. (a) 130 km/h (b) 335°22′

6. (a) Domain = {$x: x \in \Re$}
    Range = {$y: y > 0$}
   (b) R^{-1} = {$(x,y): y = 1+\log_2 y$}
   (c) -2

7. (a) $F = 44a$ (b) 110 N
   (c) 1.5 m/s²
   (d)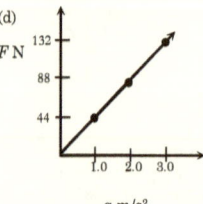

   a m/s²

8. (a) 30 cm (b) 102.8 cm²
9. (a) 312.5 cm², 70.72 cm
   (b) 5.83×10^6

10. (a) (i) $2+\sqrt{3}$ (ii) $-2-\sqrt{3}$
    (b) 0.04
11. $a = 4$ and $b = 2$
12. (a) $\frac{1}{9}$ and $\frac{1}{3}$ respectively
    (b) £1,680
13. (a) 12,220 km (b) 16,390 km
    (c) 4,170 km
14. (a) 216 (b) 120
    (c) 60   (d) 32
15. (a) (i) tzs.17 (ii) kes.0.059
    (b) kes.489.4 (c) tzs.425,000
16. (a) $p = 2, q = -4$
    (b) (i) x^2 (ii) $x-1$ (iii) $x = 1$
17. (a) $x = 0.53$ (b) 0.6349
18. (a) (i) $(x-y)(x+y-4)$
    (ii) $(2a^2+1)(1+b)$
    (b) $\frac{1}{(x+1)}$
19. (a) 200 m
    (b) 78°08′, 47°14′ and 54°38′

20. (a) 0°, 60°, 300°, 360° (b)

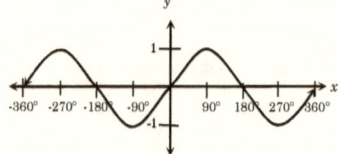

Range = {$y: -1 \le y \le 1$}

21. (a) 7.36 cm (b) 67° (c) 57°21′
22.

Dr			Cash A/c				Cr
Date	Particulars	Folio	Amount ($)	Date	Particulars	folio	Amount ($)
Jan, 1	Capital		15,000	Jan, 2	Purchases		7,000
5	Sales		13,000	9	Purchases		6,500
18	Sales		9,000	13	Carriage		150
				21	Wages		250
				31	Balance	c/d	23,100
			37,000				37,000
Feb, 1	Balance	b/d	23,100				

Dr			Capital A/c				Cr
Date	Particulars	Folio	Amount ($)	Date	Particulars	folio	Amount ($)
Jan, 31	balance	c/d	15,000	Jan, 1	Cash		15,000
			15,000				15,000
				Feb, 1	Balance	b/d	15,000

Dr				Purchases A/c			Cr
Date	Particulars	Folio	Amount ($)	Date	Particulars	folio	Amount ($)
Jan, 2	Cash		7,000	Jan,31	Balance	c/d	13,500
9	Cash		6,500				
			13,500				13,500
Feb, 1	balance	b/d	13,500				

Dr				Sales A/c			Cr
Date	Particulars	Folio	Amount ($)	Date	Particulars	folio	Amount ($)
Jan,31	balance	c/d	22,000	Jan, 5	Cash		13,000
				18	Cash		9,000
			22,000				22,000
				Feb, 1	Balance	b/d	22,000

Dr				Carriage A/c			Cr
Date	Particulars	Folio	Amount ($)	Date	Particulars	folio	Amount ($)
Jan,13	Cash		150	Jan,31	Balance	c/d	150
			150				150
Feb, 1	balance	b/d	150				

Dr				Wages A/c			Cr
Date	Particulars	Folio	Amount ($)	Date	Particulars	folio	Amount ($)
Jan,21	Cash		250	Jan, 31	Balance	c/d	250
			250				250
Feb, 1	balance	b/d	250				

23. Put 30 mathematics books and 32 physics books in a box.
24. (b)

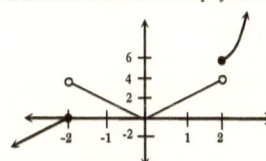

Domain = $\{x: x \in \Re\}$

Range = $\{y: y \geq 4 \text{ or } y < 2\}$

25. (a) $T = \begin{pmatrix} \frac{1}{6} & 0 \\ -\frac{1}{6} & 0 \end{pmatrix}$ (b) $2x+3y=3$ (c) $(1.5+3\sqrt{3}, 3-1.5\sqrt{3})$

Test 10

1. (a) 17.4%
   (b) Decrease by 56.9%
2. (a) 1.62 hours
   (b) $t = -5, 1 \text{ and } 3$
3. 382.52 cm^2
4. (a) 14.61 km/h (b) N79°25′E
5. (a) 125° (b) 65°
   (c) 55° (d) 100

6. (a)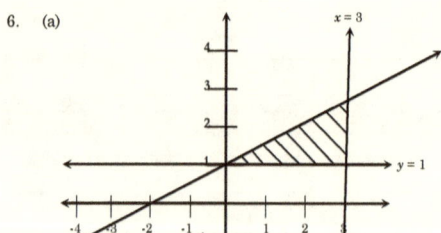

$2y = x+2$

The shaded region represents the relation.

(b) Domain = $\{x: 0 \leq x \leq 3\}$, Range = $\{y: 1 \leq y \leq 2.5\}$

7. (a) $2-\sqrt{3}$ (b) $2-\sqrt{3}$
8. 15.78 units
9. (a) 6 (b) 2 and 3 respectively
   (c) 330 and 3069 respectively.
10. (a) $114,300 (b) 2 or 0.5
11. $x = 0.5$ or $x = 2$ or $x = \dfrac{-3 \pm \sqrt{5}}{2}$
12. (a) (35°N, 160°E) (b) 4200 nm
    (c) 42 hours (d) 1:35 a.m.
13. (a) (i) 45° (ii) 8
    (b) (i) 1.93 cm
        (ii) 17.96 cm², 15.44 cm
14. (a) 12 cm and 16 cm
    (b) 100 cm², $10\sqrt{2}$ cm
15. (a) (i) $\dfrac{\sqrt{6}+2}{4}$ (ii) $\dfrac{\sqrt{6}-2}{4}$
    (iii) $2-\sqrt{3}$
    (b) $\dfrac{2+\sqrt{3}}{4}$

16. $\overline{OT}^2 = \overline{OD}^2 + \overline{DT}^2 = \overline{OC}^2 + \overline{CT}^2$ ...(i) (Pythagoras theorem)

But $\overline{CT}^2 = \overline{AT} \times \overline{BT}$ ...(ii)

Then, $\overline{DT}^2 - \overline{CT}^2 = \overline{OC}^2 - \overline{OD}^2$, from ...(i)

$\overline{DT}^2 - \overline{CT}^2 = \overline{DT}^2 - \overline{AT} \times \overline{BT} = \overline{DT}^2 - \overline{BT}(\overline{AB} + \overline{BT}) = \overline{DT}^2 - \overline{BT}^2 - \overline{BT} \times \overline{AB}$

Then,

$(\overline{DT} + \overline{BT})(\overline{DT} - \overline{BT}) - \overline{BT} \times \overline{AB} = (\overline{BD} + 2\overline{BT})(\overline{BD}) - \overline{AB} \times \overline{BT}$

$= \overline{BD}^2 + 2\overline{BT} \times \overline{BD} - \overline{AB} \times \overline{BT}$

But $\overline{BD} = \dfrac{1}{2}\overline{AB}$, (radius is $\perp$ to chord AB)

then, $\left(\dfrac{1}{2}\overline{AB}\right)^2 + 2\overline{BT} \times \dfrac{1}{2}\overline{AB} - \overline{AB} \times \overline{BT} = \overline{OC}^2 - \overline{OD}^2$

$\dfrac{1}{4}\overline{AB}^2 + \overline{BT} \times \overline{AB} - \overline{BT} \times \overline{AB} = \overline{OC}^2 - \overline{OD}^2$

$\overline{AB}^2 = 4(\overline{OC}^2 - \overline{OD}^2)$

(a) 4 cm (b) 4 cm

17. (a) shs.5,000,000 (b) 37.5%
18. (a) 8 cm, 4 cm, 18 cm and 6 cm respectively
    (b) Isosceles triangle, 15°
19. (a) $b = \sqrt[6]{\left(\dfrac{1}{c}\right)^{26}}$ (b) $x = \dfrac{b}{\sqrt[3]{a}}$ (c) $x = 3.17$
20. (a) $-1 \leq x \leq 3$ (b) 16.98

21.

Dr   Cash A/c   **Cr**

Date	Particulars	Folio	Amount (£)	Date	Particulars	folio	Amount (£)
Mar, 1	Capital		300,000	Mar, 1	Stock		20,000
3	Sales		120,000	2	Purchases		100,000
10	R. outward		5,000	6	Purchases		80,000
15	Sales		160,000	18	R. inward		8,000
26	Sales		110,000	19	Drawing		50,000
				20	Purchases		70,000
				26	Wages		8,000
				31	balance	c/d	359,000
			695,000				695,000
Apr, 1	balance	b/d	359,000				

Dr   Trading, profit and loss accounts   **Cr**

Opening stock	20,000	Sales	390,000
Purchases 25,000		Less: Return inward	8,000
Less: R. out 5,000			382,000
Net Purchases	245,000		
Cost of goods available	265,000		
Less: closing stock	25,000		
Cost of sales	240,000		
Gross profit c/d	142,000		
	382,000		382,000
Wages	8,000	Gross profit b/d	142,000
Net profit	134,000		
	142,000		142,000

Balance Sheet

Assets		Capital	300,000
Stock	25,000	Add: Net profit	134,000
Cash	359,000		434,000
		Less: Drawing	50,000
	384,000		384,000

22. (a) (i) $\dfrac{95}{192}$ (ii) $\dfrac{5}{256}$

      (iii) $\dfrac{91}{96}$ (iv) $\dfrac{17}{32}$

(b) (i) $f(x) = -3\left(x - \dfrac{2}{3}\right)^2 - \dfrac{11}{3}$

    (ii) $(\dfrac{2}{3}, -\dfrac{11}{3})$

    (iii) $x = \dfrac{2}{3}$

    (iv) Maximum value,

        $y = -\dfrac{11}{3}$

    (v) Range = $\{y: y \le -\dfrac{11}{3}\}$

23. (a) (i) 10:36 a.m. in next day

      (ii) (70°11'S, 20°W)

   (b) (i) 803.84 cm³   (ii) 563.2 cm²

24. (a) (i) $\dfrac{1}{10}\begin{pmatrix} 3 & -2 \\ -1 & 4 \end{pmatrix}$

      (ii) $x = \dfrac{13}{10}$ and $y = -\dfrac{11}{10}$

   (b) $X = \begin{pmatrix} 2 & 1 \\ 4 & -1 \end{pmatrix}$

   (c) $\dfrac{x}{10} + \dfrac{y}{5} = 1$,

      x- and y- intercepts are 10 and 5 respectively.

25. (a) 81°28'   (b) 57°15'

   (c) 430 cm³

PART 3

Examination Papers

Examination Paper 1

TIME: 3 hours

Instructions

1. This paper consists of two sections, A and B.
2. Answer **all** questions in section A and **any four** questions in section B.
3. All work done must be shown clearly.
4. Don't use electronic calculators for any question.
5. You are advised to spend no more than 2 hours in section A and using the remaining time in section B.

Section A

1. (a) If $a = 2.5 \times 10^{-5}$ and $b = 1.6 \times 10^{7}$, find the value of the expression $\dfrac{1}{\sqrt{ab}}$ in the form of $A \times 10^n$ where n is an integer; $1 \leq A < 10$.

   (b) Given that $a = 2.0\dot{3}\dot{5}$, show that a is a rational number by wring it in the form of $\dfrac{a}{b}$ where $b \neq 0$.

2. (a) Rationalize the denominator $\dfrac{\sqrt{2}+6}{2-\sqrt{2}}$, hence evaluate it if $\sqrt{2} = 1.414$.

   (b) Solve the inequality $|2x+5| > 4$, and then show the answer on the number line.

3. (a) Given the universal set µ = {All real numbers}, and its subsets A = {All natural numbers} and B = {All integers}. By describing in words, find the following sets.
   (i) A∪B
   (ii) A∩B

   (b) A crown weighs 2.5 kg was made by mixing aluminium, gold and silver in the ratio 4:3:2 respectively. The cost of 10 g of aluminium, 12 g of gold and 15 g of silver are £12, £80 and £30 respectively. Calculate the cost of the crown.

4. (a) Given that vectors $\underline{a} = 2i - 7j$, $\underline{b} = -6i + 2j$ and $\underline{c} = \underline{a} - 2\underline{b}$. Calculate
   (i) the magnitude of $\underline{c}$, and
   (ii) the direction of $\underline{c}$.
   (b) Juma drew a map of a certain town. In the town, the actual distance from point A to B is 2 km, but on his map, the distance from point A to B was 4 cm. Find
   (i) the scale of the map.
   (ii) the actual area of the town in km² if it is 7.5 cm² on the map.

5. (a) The roots of a quadratic equation $2x^2 + 3x - 5 = 0$ are p and q. Find the value of
   (i) $p^2 + q^2$ (ii) $\dfrac{1}{p} + \dfrac{1}{q}$

   (b) Solve the equation $7x^2 + 2x - 6 = 0$ by completing the square.

6. (a) Find the equation of a line which passes through the point (4,5); perpendicular to the line passing through points (3,6) and (5,8).
   (b) The midpoint of a line joining points (x,y) and (7,6), is (5,3). Find the value of x and y.

7. In a certain experiment involving a study of the relation between two parameters a and b, the parameters were found satisfy the equation $a = kb^m$. Given that, $a = 10$ when $b = 1$ and $a = 100$ when $b = 100$.

   (a) Show that $m = \dfrac{1}{2}$, and deduce the value of k.

   (b) Write the formula for $\log(a)$ in terms of $\log(b)$. Find $\log a$ when $\log b = 3$.

   (c) Draw the graph of $\log a$ against $\log b$.

8. (a) The second and tenth terms of an arithmetic progression are -1 and 63 respectively. Find the sum of the first twenty terms of the progression.
   (b) The second term of a certain geometric progression is 9 and the fourth term of the same progression is 324. If the common ratio is positive, find
   (i) the common ratio and the first term.
   (ii) the sum of the first 10 terms of the GP.

9. (a) Find the area and perimeter of the following sector of a circle.

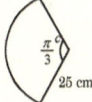

Figure 3.1.9(a).

(b) Calculate the angle $B\hat{X}C$ from the following figure if the lines $\overline{BX}$ and $\overline{CX}$ are the tangents of the circle at B and C respectively.

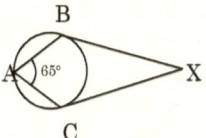

Figure 3.1.9(b).

10. From the figure 3.1.10, $\overline{AD} // \overline{BC}$, $A\hat{B}C = A\hat{C}D = \overline{AD} = 4$ cm and $\overline{BC} = 9$ cm.
   (a) Calculate the length $\overline{AC}$
   (b) Prove that $\dfrac{AD}{BC} = \left(\dfrac{CD}{AB}\right)^2$

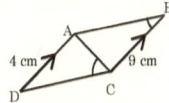

Figure 3.1.10.

Section B

11. The table below shows the number of units per kilogram of protein, fats and carbohydrates contained in two types of food and the minimum required units of each per day.

	Food X	Food Y	Number of units required
Proteins	4	1	20
Fats	2	1	16
Carbohydrates	1	1	12
Cost	shs.4000	shs.3000	

How many kilograms of each type of food should be bought in order to minimize the cost?

12. The following frequency distribution table shows the marks scored by pupils in a certain test.

Marks	0–10	10–20	20–30	30–40	40–50	50–60	60–70
Frequency	8	14	24	26	21	15	2

(a) Draw a histogram, and a frequency polygon on the same axis.
(b) Calculate the mean mark by using the assumed mean equal to 25.
(c) Determine the modal class and the mode mark.
(d) Determine the median class and median mark.

13. Below is a hemispherical bowl which is closed on the top. The bowl has the radius of 21 cm.

Figure 3.1.13.

(a) Calculate the total surface area, and the amount of water in litres that can fill the bowl.
(b) If the bowl is used to fill the water in the tank of height 2 m; and the area of the base 0.4851 m², how many bowls can be used to fill the tank? (Use $\pi = 22/7$).

14. On 1st January 2010, Mr. Omar commenced a business with capital in cash $600,000.

Jan 2, Bought goods for cash	$150,000
3, Sales for cash	$100,000
5, Purchased goods for cash	$50,000
10, Paid transport charges	$5,000
13, Sold goods for cash	$200,000
18, Bought goods for cash	$85,000
22, Cash sales	$320,000
30, Paid wages	$3,500

Balance the cash account, take out the trial balance and prepare the trading and profit and loss accounts. Stock on hand at 31st January $9,500.

15. (a) A vector $\mathbf{a} = 2\mathbf{i}+3\mathbf{j}$ reflected along the line $y = x$ followed by the rotation of 90° anticlockwise direction about the origin. Find the final image of the vector $\mathbf{a}$, then find the magnitude of the image.

(b) If the inverse of a matrix $\begin{pmatrix} 2 & 2a \\ a & 6 \end{pmatrix}$ is $\begin{pmatrix} 3b & -1 \\ -0.5 & b \end{pmatrix}$, find the values of a and b.

16. (a) The function f is defined as $f(x) = \begin{cases} x & if \quad x \geq 3 \\ x^2 & if \quad 0 < x < 3 \\ |x| & if \quad x < 0 \end{cases}$

(i) Sketch the graph of f.
(ii) State the domain and range.
(iii) Find $f(-\pi)$ and $f(0)$
(iv) State whether f is one-to-one function.

(b) A bag contains three black balls, five white balls and two red balls. Two balls are drawn randomly from a bag one after another without replacement. What is the probability of drawing two balls
(i) both being white?
(ii) both being red?
(iii) of different colours?

Examination Paper 2

TIME: 3hours

Instructions

1. This paper consists of two sections A and B.
2. Answer **all** questions in section A and **any four** questions in section B.
3. All work done must be shown clearly.
4. Don't use electronic calculators for any question.
5. You are advised to spend no more than 2 hours in section A and using the remaining time in section B.

Section A

1. (a) Use mathematical tables to evaluate $\left(\sqrt{\dfrac{25.47}{0.0873}}\right)^3$

   (b) Find the product of the LCM and GCF of numbers 24, 30 and 42.

2. (a) James bought a car for £50,000 four years ago. The car depreciates its value by 10% each year. Calculate the current cost of the car.

   (b) Make P the subject of the formula from
   $$P = \sqrt{\dfrac{(PQ)^3 - R}{2P}}$$

3. (a) You have given that the universal set $\cup$ = {1,2,3,4...10}, and its subsets are A = {2,4,6,8,10}, B = {3,6,9}, and C = {3,4,5,6,7}. Represent the information in Venn diagram, then find n(A∪B) and n(A∩B∩C).

   (b) Three people shared $18,000/= in the ratio 2:3:4. How much will each get?

4. (a) If vectors $\underline{r} = 2i - j$, $\underline{s} = 2j+i$, and $\underline{t} = -3i - 4j$, find
      (i) $|\underline{r}+2\underline{s} - 3\underline{t}|$
      (ii) the value of scalars x and y if $x\underline{t} - y\underline{s} = 2\underline{r}$

   (b) Find the magnitude of the resultant force between the following forces which acting at point O.

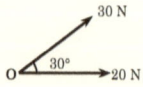

Figure 3.2.4.

5. (a) Find the value of $528.25^2 - 471.75^2$
   (b) Solve the following simultaneous equations:

$$\begin{cases} \dfrac{1}{2}x + \dfrac{1}{3}y = 4 \\ \dfrac{3}{4}x - \dfrac{3}{2}y = -6 \end{cases}$$

6. (a) Find the equation of a line which passes through points (-1,-1) and (4,9).
   (b) From the equation in (a) above, find the point on it, equidistant from the points (6,2) and (5,1).

7. Consider the currencies' exchange below

$1 = tzs.2,300
£1 = tzs.3,100
¥1 = tzs.360

Convert (a) ¥5000 into British Pound (£)
        (b) $2580 into Chinese Yuan (¥)
        (c) £4000 into US Dollar ($)

8. (a) Find the sum of the first hundred integers; excluding the multiples of four and seven.
   (b) What is the sum of the infinity series 100+75+56.25+...?

9. (a) Find the shaded area common to the intersecting circles below if PQ = 20 cm. P and Q are the centres of the circles.

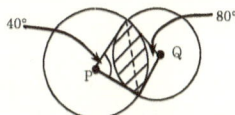

Figure 3.2.9(a).

(b) Find the relation satisfying the shaded region on the figure 3.2.9(b) below.

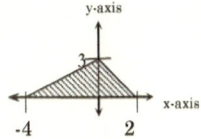

Figure 3.2.9(b).

10. From figure 3.2.10 below, ABCD is a trapezium in which AB is parallel to CD; AB = $3x$; AD = $4x$; and CD = $5x+5$ cm. The area of the trapezium is 84 cm². Find
   (a) the value of x (b) the length BE
   (c) the length EC   (d) the length BC

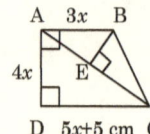

Figure 3.2.10.

Section B

11. (a) Maximize and minimize $f(x,y) = 4y+2x$ subject to
    $x \geq 0$, $y \geq 0$, $2x+y \geq 11$, $x+3y \geq 13$ and $3x+2y \leq 19$
    (b) Solve the equation $\dfrac{x+3}{1-x} = \dfrac{2x+1}{2x-1}$ by completing the square.

12. The table below represents the distribution of people with their monthly salaries in a certain company.

Salaries × 1000 ($)	1.5–1.9	2.0–2.4	2.5–2.9	3.0–3.4	3.5–3.9
Number of people	4	15	12	6	3

    By estimating to the nearest units,
    (a) calculate the mean monthly salary to each person.
    (b) find the mode monthly salary.
    (c) determine the median class and median monthly salary.

13. On the figure 3.2.13 below, AF = 10 cm, AD = 4 cm and AB = 6 cm. Calculate
    (a) the length BF,
    (b) the length AG,
    (c) the angle between the line AF and the line EF, and
    (d) the angle between the line AG and the plane EFGH.

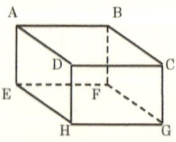

Figure 3.2.13.

14. Enter the following transactions in the cash account, balance off at the end and bring the balance, and hence construct a trial balance.

   April, 2000 1st – Cash in hand £35,000

               2nd – Paid office expenses for cash £7,500

               4th – Purchased goods for cash £8,000

               5th – Sold goods for cash £15,000

               8th – Paid carriage for cash £8,500

               12th – Purchased goods for cash £10,000

               13th – Paid garage expenses £6,000

               18th – Sold goods for cash £18,000

               21st – Purchased goods for cash £6,000

               23rd – Sold goods for cash £28,000

               25th – Paid wages for cash £1,500

15. (a) The matrix M transforms the point (2,-3) to the point (-5,18), and the point (3,3) to the point (15,-3). Find

   (i) the matrix M.

   (ii) the image of point (4,5) after transformed by the matrix M.

   (b) Use matrices to solve the simultaneous equations below.

   $$2x - 5 = y \text{ and } 5y + x = 8$$

16. (a) For what value of 'a' would the polynomial function $f(x) = 2x^3 + ax^2 - 4x - 3$ leave the remainder 17 when divided by $(x-2)$?

   (b) If $f(x) = x - 5$ and $g(x) = 2x + 3$, determine

   (i) fg (ii) $f+g$ (iii) $f_o g$

   (c) The probability that a woman will go to out tomorrow, and the probability that she will take her son with her are 0.5 and 0.4 respectively. If she won't go out tomorrow, the probability that her son will go is 0.3. Find the probability that

   (i) the women will go out without her son.

   (ii) her son will not go out tomorrow.

Examination Paper 3

TIME: 3 hours

Instructions

1. This paper consists of two sections A and B.
2. Answer **all** questions in section A and **any four** questions in section B.
3. All work done must be shown clearly.
4. Don't use electronic calculators for any question.
5. You are advised to spend no more than 2 hours in section A and using the remaining time in section B.

Section A

1. (a) Simplify $\dfrac{\frac{1}{2} + \frac{5}{8} \div \frac{10}{24}}{\frac{7}{5} + \frac{9}{8}}$

   (b) The LCM and GCF of numbers x, 420 and 180 are 2520 and 60 respectively. Find the value(s) of x.

2. (a) If $\log a = 1.4314$ and $\log b = \bar{2}.3221$, find the value of $\dfrac{a}{b}$

   (b) Simplify the following expression:

   $$\frac{1}{x^2+5x+6} + \frac{1}{x^2+3x+2} + \frac{1}{x^2+7x+12}$$

3. (a) If A_x = {Factors of x} and B_x = {Multiples of x}, list the elements of the following sets.

   (i) $A_{24} \cap B_2$ (ii) $A_{12} \cup B_6$

   (b) In a class of 44 students, 17 of them like Kiswahili, 17 like English, 17 like Arabic and 2 students like all the three languages. It is also found that 7 students like both Kiswahili and English, 5 students like both Kiswahili and Arabic and 6 students like English only. Represent the information in Venn diagram and use it to find the number of students who

   (i) like Kiswahili only.

   (ii) like both English and Arabic.

   (iii) like Kiswahili or Arabic

   (iv) do not like any of these languages.

4. Given the points P = (2,-1), Q = (-3,5) and R = (0,4), and that $\mathbf{r} = \overrightarrow{PQ} + 3\overrightarrow{RP} - 2\overrightarrow{RQ}$.

   (a) Find $\mathbf{r}$ as a coordinate.

   (b) Determine $|\mathbf{r}|$.

   (c) Find the direction of $\mathbf{r}$.

5. (a) Find the value of a from figure 3.3.5 below.

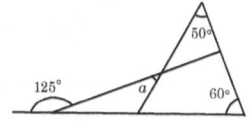

Figure 3.3.5.

(b) Solve for θ from the equation $1+\cos\theta = 2\sin^2\theta$ for $0° \leq \theta \leq 360°$

6. (a) Calculate the area of the trapezium PQRS below.

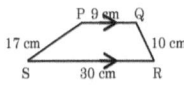

Figure 3.3.6.

(b) The area of a triangle whose sides are 3 cm, 5 cm and 6 cm is equal to the area of a circle whose diameter is d. Find the value of d.

7. (a) Find the equation of a perpendicular bisector of a line segment whose end points are (3,7) and (-5,3).
   (b) What the value of 'a' will make the line which passes through the points (a,7) and (1,1) perpendicular to the line which passes through the points (a,-5) and (4,a)?. Find the point of intersection of these two lines.

8. (a) Given that P is partly constant and partly varies as the square of Q. Given that P = 4 when Q = 10 and P = 10 when Q = 30. Find
   (i) the equation connecting P and Q.
   (ii) P when Q = 8.

   (b) If $p = \dfrac{1}{1-a}$ and $a = \dfrac{1}{1-b}$, show that $b = \dfrac{1}{1-p}$

9. (a) Find the value of θ from the circle below if O is the centre, $\angle AFE = 120°$, $\angle CBD = 40°$, $\angle ACB = 30°$ and $\angle ABE = \theta$

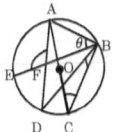

Figure 3.3.9.

(b) Prove that the angles in alternate segment of a circle are equal.

10. (a) The area of the forest as seen on the map is 2.5 cm² while its actual area is 0.1 km². Find
    (i) the scale of the map.
    (ii) the distance of the road on the map if it is 20.8 km on the ground.
  (b) The second, fourth and eighth terms of an arithmetic progression forms the first three terms of a geometric progression. If the sum of the third and fifth terms of the progression is 80, find the sum of the first ten terms of the progression.

Section B

11. A blacksmith has 0.4 tons of iron and 35 litres of paints to make windows and doors for the house. To make a window, he needs 10 kg of iron and 1 litre of paints. To make a door, he needs 25 kg of iron and 2 litres of paints. If the cost of each window is shs.45,000 and the cost of each door is shs.100,000, how many windows and doors should he make in order to maximize his profit, and what is the profit?

12. In a certain random experiment, Ali rolled an ordinary die 112 times. He recorded the scores he got on the table below.

Scores	1	2	3	4	5	6
Frequency	38	13	14	8	20	19

  (a) Calculate (i) the mean score,
               (ii) the median score, and
               (iii) the mode score.
  (b) With regard to the experiment, find the probability that
    (i) the score is greater than 3.
    (ii) the score is exactly 4.

13. (a) Two ships X and Y leaves ports A(20°N, 32°E) and B(15°S, 32°E) respectively and sails towards each other along the great circle. If the speed of the ships are 20 knots and 30 knots respectively,
    (i) how long would it take for the ships to meet?
    (ii) at what point will they meet?
  (b) A plane set off on a course of 100° with an airspeed of 200 knots. Its track due to the wind was 180° with ground speed of 150 knots. Calculate the speed and direction of the wind.

14. The following information was written from a certain businessman books.

   Sales: $30,000
   Purchases: 60% of sales
   Closing stock: 50% of purchases
   Net loss: $1,500
   Opening stock: 50% of gross profit.

   Find
   (a) the gross profit,
   (b) the opening stock,
   (c) the cost of goods available for sale,
   (d) the cost of sales, and
   (e) the expenses.

15. (a) The matrix X transforms the square PQRS with vertices P(0,0), Q(0,3), R(3,3) and S(3,0) on vertices P'(0,0), Q'(15,24), R'(21,27) and S'(6,3) respectively.
   (i) Find the matrix X.
   (ii) Determine the area of P'Q'R'S'.

   (b) If $A = \begin{pmatrix} 1 & 2 \\ 2 & 8 \end{pmatrix}$, find the product of BA if the product $AB = \begin{pmatrix} 16 & 4 \\ 24 & 24 \end{pmatrix}$.

16. (a) Given that $f(x) = 2^x + 4$.
   (i) Find the domain and range of f.
   (ii) Determine $f^{-1}(x)$.

   (b) Draw the graph of $g(x) = -x^2 + 2$. From the graph, state the domain and range.

   (c) Bag A contain six red balls and four white balls, and bag B contain three red balls and nine white balls. If a bag is selected randomly and then a ball drawn at random, find the probability that a ball is
   (i) red
   (ii) white

Examination Paper 4

TIME: 3 hours

Instructions

1. This paper consists of two sections A and B.
2. Answer **all** questions in section A and **any four** questions in section B.
3. All work done must be shown clearly.
4. Don't use electronic calculators for any question.
5. You are advised to spend no more than 2 hours in section A and using the remaining time in section B.

SECTION A

1. (a) By using mathematical tables, evaluate

$$\sqrt{\frac{11.35}{7.68} + 1.13^3}$$

   (b) James bought a radio which cost him shs.40,000 and later he decided to sell it with profit, for shs.x. But while he was selling it, after negotiation with his customer, he reduced his selling price by 15% and he was remained with only 27.5% profit of the cost of the radio. Calculate
   (i) the value of x.
   (ii) the percentage profit before the reduction of selling price.

2. (a) Find x and y from figure 3.4.2 below.

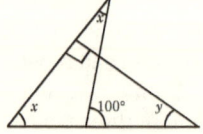

Figure 3.4.2.

   (b) Simplify $\dfrac{\sqrt{2}+8}{4-\sqrt{8}} \div \dfrac{5\sqrt{2}}{\sqrt{2}+6}$

3. (a) In a certain school, 10 students play football, 12 play volleyball, 4 play both football and volleyball, 10 play netball, 15 play none of the sports and 3 play both volleyball and netball. Those who play football do not play netball.
   (i) Represent the information in Venn diagram.
   (ii) How many students do not play any game among these three games?
   (iii) How many students are in the school?

   (b) Solve the inequality $\dfrac{4x-3}{2} \geq \dfrac{3}{2}x - \dfrac{1}{4}$

4. (a) Given the points P(2,5), Q(-3,6) and R(-1,4). Find
   (i) the vectors $\overrightarrow{QP}$ and $\overrightarrow{QR}$.
   (ii) the magnitude of vector $\overrightarrow{RQ} - \overrightarrow{QP}$.

   (b) Two buses left station A at the same time. The first bus was moving on bearing of 055° at a speed of 80 km/h and the second bus was moving on the bearing of 300° at a speed of 95 km/h. How far apart are they after 3.5 hours?

5. (a) The average weight of 20 students is 59.05 kg and the average weight of 13 students among them is 58 kg. Calculate the average weight of the students left.

   (b) When the expression $ax^3 + 3x^2 - bx + 6$ is divided by $x - 2$, it leaves the same remainder as it divided by $x+1$. Show that $b = 3a + 3$.

6. (a) If x varies as y and inversely as z, and $x = 2$ when $y = 4$ and $z = 6$. Find
   (i) The equation that connect x, y and z.
   (ii) x when $y = 3$ and $z = 5$.

   (b) If the height of a cone decreases by 15% and its volume remains constant, what must the radius of the base increases by?

7. (a) Two vertices of an isosceles triangle ABC with AB = AC are A(12,3) and B(4,-3). The equation of the line BC is $x+y = 1$, and D is the midpoint of $\overline{BC}$. Find
   (i) the points C and D.
   (ii) the equation of the line AD.

   (b) What is the shortest distance from point P(2,-4) to the line $y = -\dfrac{2}{5}x + 5$?

8. (a) How many terms must be added to the sequence 7, 15, 23,... to make the sum equal to 945?
   (b) The arithmetic and geometric means of two numbers are 10 and 8 respectively. Find the numbers.

9. (a) If $\cos\theta = \dfrac{8}{17}$, without using tables evaluate

   (i) $2\sin^2\theta + 3\cos^2\theta$ (ii) $\dfrac{\sin\theta + \cos\theta}{\cos 2\theta}$

   (b) If θ and α are complementary angles, show that $\tan\alpha = \dfrac{\cos\theta}{\sin\theta}$

10. From figure 3.4.10 below, given that AB = 12 cm, A'B' = 4.5 cm, and B'C = 6 cm. Calculate

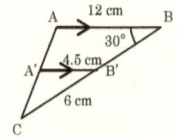

Figure 3.4.10.

   (a) the length BB',
   (b) the length AC and A'C, and
   (c) the area of triangle AB'C'.

Section B

11. The manager of EDS company has shs.1,375,000 and he decided to buy some televisions and radios for shs.45,000 and shs.35,000 each respectively in order to sell. Each television occupies 4 m² of space on the floor, and each radio occupies 2 m² of space on the floor. He has only the room of 100 m² space of floor to keep them. Each device cannot be put above the other. Also, each television can take 2 minutes to put in the room and each radio takes 6 minutes to put in the room. If the manager needs to gain a profit of 100% when he sell the items but he doesn't want take the maximum time to put them in the room, how many televisions and radios should he buy in order to get the maximum profit?

12. (a) Given the data 2, 5, 7, 6, 4, 7, 5, 3, 6, 5. Find
    (i) the mean   (ii) the median   (iii) the mode.
   (b) Mr. Mick spent shs.252,000 on six types of goods. The prices of the goods were recorded on the pie chart as shown below.

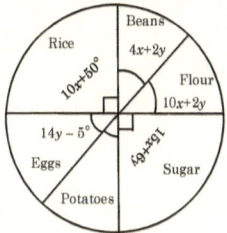

Figure 3.4.12.

   (i) Find the value of x and y.
   (ii) How much did Mr. Mick spend on potatoes?

13. From the right pyramid VABCD below, AB = 8 cm, BC = 10 cm and VC = 14 cm.

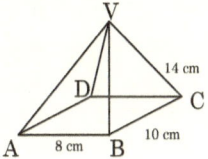

Figure 3.4.13

Calculate
(a) the height of the pyramid,
(b) the angle that VA makes with the base ABCD,
(c) the angle between the plane VDC and the base, and
(d) the volume of the pyramid.

14. Record the following transactions in the Mr. Msuba's ledger on double entry principles.

    September 1, Msuba commenced business with capital in cash £100,000
    Sept 2, purchased goods for cash £55,000
        5, sold goods on credit to Madevu £25,000
        10, purchased goods for cash £40,000
        15, sold goods for cash £82,000
        23, received cash from Madevu £15,000

    Balance the accounts.

15. (a) The following is a geometric progression.

$$\begin{pmatrix} \frac{12}{5} & -2 \\ 4 & -\frac{9}{5} \end{pmatrix} + \begin{pmatrix} -\frac{18}{5} & 3 \\ -6 & \frac{27}{10} \end{pmatrix} + \ldots + \begin{pmatrix} \frac{2187}{80} & -\frac{729}{32} \\ \frac{729}{16} & -\frac{6561}{320} \end{pmatrix}$$

   (i) Find the number of terms in the series.
   (ii) Determine the sum of the series.

(b) A linear transformation maps the point (x,y) into the point (x',y') such that

$$\begin{pmatrix} x' \\ y' \end{pmatrix} = \begin{pmatrix} 3 & 2 \\ -1 & 4 \end{pmatrix} \begin{pmatrix} x \\ y \end{pmatrix} + \begin{pmatrix} 3 \\ -6 \end{pmatrix}.$$

Find
   (i) the image of point $(2,5)$ under this transformation.
   (ii) the point whose image is $(-1,3)$.

16. (a) Given that $f(x) = 3x^2 - 4x + 7$. Find
   (i) the turning point of the graph of $f(x)$,
   (ii) the line of symmetry, and
   (iii) the maximum or minimum value of f.

(b) The independent events A, B and C are such that $P(C) = \frac{1}{4}$, $P(B \cup C) = \frac{5}{8}$ and $P(A \cup B) = \frac{4}{5}$. Find

   (i) P(B)   (ii) P(A)

(c) The events A and B are said to be independent. Consider the following tree diagram.

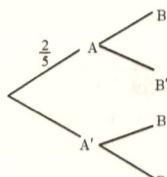

Figure 3.4.16

If $P(A \cap B) = 0.15$, find
   (i) the probability of occurring A' and B.
   (ii) the probability of occurring A or B.

Examination Paper 5

TIME: 3 hours

Instructions

1. This paper consists of two sections A and B.
2. Answer **all** questions in section A and **any four** questions in section B.
3. All work done must be shown clearly.
4. Don't use electronic calculators for any question.
5. You are advised to spend no more than 2 hours in section A and using the remaining time in section B.

Section A

1. (a) Use logarithm tables to evaluate
$$\frac{0.0752 - 0.05832}{\sqrt{0.0752} + \sqrt[3]{0.05832}}$$

   (b) Write the answer in (a) above in standard form into two significant figures.

2. (a) Given that $2\log_a\left(\frac{1}{b}\right) = 8$, find $-2\log_b\left(\frac{1}{a}\right)$

   (b) Without using mathematical tables, show that
$$\frac{2^{100}}{10^{50}} = (0.16)^{25}$$

   (c) Factorize completely the expression $(a - b)^3 + b - a$.

3. (a) If sets A and B are the subsets of a universal set $\cup$ and that $\cup$ = {Whole numbers less than 20}, A = {Odd numbers}, and B = {Factors of 60}. Represent the given sets on Venn diagram and determine
   (i) n(A∩B)   (ii) n(A∪B)'

   (b) In the village of 100 men, 40 are farmers, 45 are fishermen only and 10 are both farmers and fishermen. Find the number of men who are
   (i) farmers only.
   (ii) Fishermen.
   (iii) neither farmers nor fishermen in the village.

4. Given that $\underline{u} = a\boldsymbol{i}+b\boldsymbol{j}$, $\underline{v} = 2a\boldsymbol{i} - b\boldsymbol{j}$, $|\underline{u} - \underline{v}| = 5$, $T = \begin{pmatrix} 1 & 2 \\ 2 & 3 \end{pmatrix}$, and $|T(\underline{u}+\underline{v})| = \sqrt{405}$. Calculate
   (a) the value of a and b, for all a, b ≥ 0.
   (b) $T(3\underline{u}+2\underline{v})$.
   (c) $|T(\underline{p})|$ if $\underline{p} = 2\underline{u} - 3\underline{v}$.

5. (a) Given that x is directly proportional to y. The sum of the values of x when $y = 2$ and $y = 4$ is 18.
   (i) Write the equation which show the relation between x and y.
   (ii) Find x when $y = 8.2$
   (b) Ali whose height is 1.5 m standing on a top of a tower observes a car due east at an angle of depression of 30°. When the car moves 30 m towards the foot of the tower, the angle of depression is 60°.
   (i) Calculate the height of the tower.
   (ii) If the speed of the car is 1.5 m/s, find the time taken to reach the foot of the tower from the initial position.

6. (a) The vertices of triangle PQR are P(2,7), Q(5,3) and R(a,b). Point T divides the line QR internally in the ratio 1:2, and that PR > PQ, and ∠PTR = 90°. The equation of the line PT is $3x+y = 13$. Find
   (i) the points T and R.
   (ii) the length PR.
   (b) Given that α and β are the roots of a quadratic equation $5x^2 - 7x = 10$. Find
   (i) the value of $\alpha+\beta$ and $\alpha\beta$.
   (ii) the equation whose roots are $\dfrac{\alpha}{\beta}$ and $\dfrac{\beta}{\alpha}$.

7. (a) Calculate the length AB on a triangle ABC if ∠A = 50°, AC = 10 cm and BC = 8 cm.
   (b) From the triangle below, prove that $\tan\theta = \dfrac{c\sqrt{3}}{2a-c}$

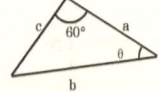

Figure 3.5.7.

8. (a) The difference between the eighth and sixth terms and the difference between the tenth and sixth terms of the same geometric progression whose common ratio is a positive integer are 168 and 1680 respectively. Find the common ratio of the progression.
   (b) How long would it take for the amount of money to double when invested in a bank at 7.15% compounded annually?

9. (a) From the circle ABCQ, PQR is a tangent at Q, ∠BQC = 20°, AQ = AC, and BQ is the diameter. Calculate
   (i) ∠ACB   (ii) ∠AQP

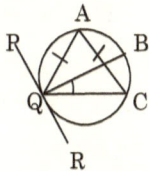

Figure 3.5.9.

   (b) A ship of speed 200 knots starts at (20°S, 30°E) and sails due south for 10 hours, then due west for 6 hours. Find the latitude and longitude of its final position.

10. (a) A tube has its outer and inner diameters of 21 mm and 10.5 mm respectively. If the tube is 2 m long, find the volume in cm³ of the material used to make the tube. (Use $\pi = \dfrac{22}{7}$)

    (b) From the diagram below, APC and ASB are the tangents at P and S respectively. The minor arc PQ is equal to the minor arc RS and O is the centre of the circle. Prove that
    (i) △APO ≡ △ASO   (ii) △POC ≡ △SOB   (iii) AC = AB

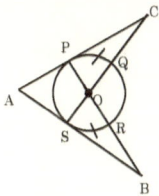

Figure 3.5.10.

Section B

11. (a) Minimize $F(x,y) = 3x+3y$ subject to the following inequalities:

    $2x+y \geq 9$, $3x+4y \geq 26$, $x \geq 0$ and $y \geq 0$

    (b) Draw the graph which satisfy the inequalities below.

    $y \geq 0$, $2x+y < 6$ and $y - x < 3$

12. The following are the recorded scores obtained by 40 students in a basic mathematics examination.

35	55	27	38	44	60	71	49	69	56
48	56	49	53	45	18	19	52	27	08
27	64	80	38	57	26	24	47	47	68
56	32	73	46	23	12	56	36	07	29

    (a) Group the scores in the class intervals of 0–20, 20–40, 40–60, etc.
    (b) Draw a histogram to represent the data and use it to estimate the mode.
    (c) Draw a cumulative frequency curve and use it to estimate the median.

13. (a) Given that $-3 \leq x \leq 2$ and $-6 \leq y \leq 5$. Calculate
    (i) the largest possible value of xy.
    (ii) the smallest possible value of xy.
    (b) The volume and surface area of a rectangular prism are 192 cm^3 and 208 cm^2. If the height of the prism is 4 cm, find
    (i) the other two non-parallel sides.
    (ii) the lengths of the diagonals of the prism.

14. The following trial balance was extracted from the books of Mr. Omar on 30th Nov, 2009.

Account	Dr ($)	Cr ($)
Capital	-	40,000
Sales	-	82,000
Purchases	48,000	-
Drawing	4,000	-
Land and building	10,000	-
Furniture	9,800	-
Debtors	8,800	-
Creditors	-	4,600
Cash	30,500	-
Opening stock	3,250	-
Insurance	5,470	-
Wages	6,580	-
Loan from bank	-	6,000
Bad debts	6,200	-
	132,600	132,600

The closing stock was $4,800. You are required to prepare
(a) the trading, and profit and loss accounts, and
(b) the balance sheet
as on 30th Nov, 2009.

15. (a) The image of point (7,3) after reflected along the line L was (5,9). Find
    (i) the equation of line L.
    (ii) the image of (-2,1) after reflected along the line L.

   (b) If matrix $A = \begin{pmatrix} 1 & 2 \\ 4 & 3 \end{pmatrix}$ and $B = \begin{pmatrix} 2 & 0 \\ -1 & 3 \end{pmatrix}$
    (i) Find A^2 and B^2.
    (ii) Verify that $A^2 - B^2 \neq (A+B)(A-B)$.

16. (a) Write $f(x) = 6 - 2x^2 - 4x$ in the form of $f(x) = a(x-b)^2 + c$.
   (b) If the domain of a function $f(x) = 2x^2 + 5x - 3$ is $-3 \leq x < 10$, find the range of f.
   (c) The probability that Juma will go to school tomorrow is 0.89, and that of Fatma is 0.91 respectively. Find the probability that tomorrow,
    (i) both of them will go to school.
    (ii) at least one of them will go to school.
    (iii) no one will go to school.

Examination Paper 6

TIME: 3 hours

Instructions

1. This paper consists of two sections A and B.
2. Answer **all** questions in section A and **any four** questions in section B.
3. All work done must be shown clearly.
4. Don't use electronic calculators for any question.
5. You are advised to spend no more than 2 hours in section A and using the remaining time in section B.

Section A

1. (a) Three bottles have the volumes of 150 cm³, 250 cm³ and 300 cm³. Find the volume of the largest container which can be used to fill each of the bottles an exact number of times.
   (b) Use mathematical tables to evaluate the following and write the answer in the form of $a \times 10^n$ where n is any integer and $0 \leq a < 10$, correct into two significant figures.

   $$\frac{52.87 \times \sqrt{32.3}}{(45.23)^2}$$

2. (a) Find the value of a if $\log \sqrt[4]{a} = \overline{3}.2723$
   (b) Solve for a from the logarithmic equation

   $$2a^{\log 2a} = 200a$$

3. (a) The lengths of minor and major arcs of a circle are in the ratio 3:5. If the area of the circle is 49π cm², calculate the lengths of the arcs. (Use $\pi = {}^{22}\!/_{7}$)
   (b) The surface area of a solid cylinder whose radius is 14 cm, is 2112 cm².
      (i) Find the volume of the cylinder.
      (ii) If the cylinder is to be melted down and recast into a solid sphere, calculate the radius of the sphere into three significant figures.

4. (a) If a binary operation * over real numbers a and b is defined as, $a*b = 2a+5b$, find
   (i) $(1*-2)*3$
   (ii) x if $(2*x)*(x*2) = 78$

   (b) Given that $f(x) = 2x^2 + 5$ and $g(x) = x - 2$. Using a long division, find the quotient of $\dfrac{f(x)}{g(x)}$ and state the remainder.

5. (a) Sets A and B are defined as follows:

   $A = \{x : -3 \leq x < 4\}$ and $B = \{x : -1 < x \leq 7\}$.

   Represent the following sets on the number line.

   (i) A∪B
   (ii) A∩B

   (b) In a class of 40 students, 23 are boys. Given that 24 students play football and 22 play volleyball. Among the boys, 2 play neither of the game while among the girls, 8 play football, 5 play volleyball and 3 play both of the games. Represent the information on Venn diagram, then find the number of
   (i) students who play either football or volleyball.
   (ii) students who play neither of the games.
   (iii) boys playing both of the games.

6. (a) A is the point (5,-2) and B is the point (-3,4). Use the position vectors to find the magnitude of vectors
   (i) $\overrightarrow{AB}$
   (ii) $\overrightarrow{BA}$

   (b) On the figure 3.6.6 below, $\overrightarrow{PQ} = \underline{x}$ and $\overrightarrow{QR} = \underline{y}$; L is the midpoint of PR, and $\overrightarrow{QR} = 3\overrightarrow{QS}$. Write the following in terms of $\underline{x}$ and $\underline{y}$.

   (i) $\overrightarrow{SP}$
   (ii) $\overrightarrow{SL}$
   (iii) $\overrightarrow{PL}$

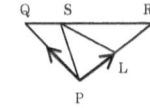

Figure 3.6.6.

7. The line segments AB and CD of equations $3x+y = 10$ and $7x+y = 26$ meet at point X where the line PQ which is perpendicular and parallel to the line BD and the line segment TS respectively pass through. If point S is on the line BD, and that points S and T are (-3,-1) and (2,-6) respectively, find
   (a) the equations of lines PQ and BD.
   (b) the points B and D.
   (c) the perimeter of triangle XBD.

8. (a) Two parallel chords of a circle, of lengths 16 cm and 30 cm, lie on opposite sides of the centre. If the radius of the circle is 17 cm, find the distance between the chords.
   (b) From figure 3.6.8 below, O is the centre of the circle. The length of the minor arc AB = 18 cm and the radius is 10 cm. Calculate
      (i) the length of the chord AB
      (ii) ∠BAC
      (iii) the area of triangle ABC.

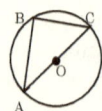

Figure 3.6.8.

9. (a) The sum of the first n terms of a sequence is given by $S_n = 2n^2 + 3n$. Find the first three terms and the general formula of the sequence.
   (b) The sum of the first two terms of a geometric progression is 16 and the sum of the next two terms is 144. Find the possible values of the common ratio and the first term of the progession.

10. The figure 3.6.10 below is a parallelogram. Given that $\overline{AE}$ = 4 cm, $\overline{FC}$ = 6 cm and $\overline{DC}$ = 8 cm.

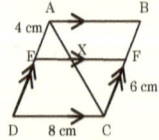

Figure 3.6.10.

(a) Write down a pair of congruent triangles with reasons.
(b) With reasons, which triangles are similar to ΔXFC?
(c) Find $\overline{EX}$ and $\overline{XF}$.

Examination Papers

Section B

11. (a) Draw on the same axis the graphs of $y = \sin 2x$ and $y = \cos x$ if $0° \leq x \leq 180°$. Hence, use your graph to solve the equation $\sin 2x = \cos x$.

   (b) Given that x and y are the first and second positive integers respectively. Three times the second integer is not more than the difference between 12 and four times the first integer. The difference of the second from the first integer is less than 1. Furthermore, the sum of thrice the first integer and the second integer is more than 4. Draw the graph to represent the information above, and then find the two integers.

12. The scores of 42 students from a geography exam were recorded on the table below.

Class mark	10	30	50	70	90
Frequency	5	x	$3y$	$y+8$	2

The mode score were found to be 56. Find

(a) the value of x and y,
(b) the mean mark, and
(c) the median.

13. (a) Calculate the total surface area of the figure 3.6.13(a) below if the radius of the base is 7 cm.

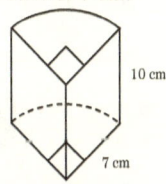

Figure 3.6.13(a).

   (b) The part of the cone of height 20 cm is removed out and leaving the shape of the frustum which has the top and bottom radii of 6 cm and 10 cm respectively as shown on the Figure 3.6.13(b) below. Calculate
   (i) the volume of a part of the cone removed.
   (ii) the volume of a part left (frustum).

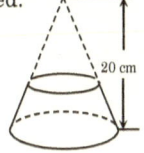

Figure 3.6.13(b).

14. The following trial balance was extracted from the books of UMOJA group on 31st July, 2002. You are required to prepare
    (a) the trading, and profit and loss accounts, and
    (b) the balance sheet
    as on 31st July, 2002.

ACCOUNT	Dr ($)	Cr ($)
Capital	-	75,000
Cash	72,550	-
Opening stock	8,700	-
Purchases	94,300	-
Sales	-	125,550
Wages	14,000	-
Rent	11,000	-
TOTAL	200,550	200,550

Closing stock was $8,000

15. (a) The equation of the line $y = 2x+3$ is transformed by transformation

$$M = \begin{cases} x' = x + y \\ y' = 2x - y \end{cases}$$

Find: (i) The matrix M (ii) The image of the equation.

(b) The translation T with vector $\begin{pmatrix} 2 \\ -5 \end{pmatrix}$ takes the point A(b,-4) onto the point B(3,-a^2). Find a and b. Also find where T takes the point C(-3,3).

(c) Given matrices $A = \begin{pmatrix} 1 & 3 \\ -2 & 0 \end{pmatrix}$ and $B = \begin{pmatrix} 5 & -2 \\ 1 & 4 \end{pmatrix}$.

Find (i) $(A+B)^2$ (ii) $2B^{-1} - A^{-1}$

16. (a) Given that $f(x) = 0.5^x$
    (i) Find the domain and range of f.
    (ii) What the value of x will give $f(x) = 16$?
  (b) The population of a certain species of fishes is decreasing at 2.5% each year. Now, the population is 3 million.
    (i) What was the population 4 years ago?
    (ii) When will the population reach 1.5 million?
  (c) (i) How many whole numbers between 100 and 10000 can be formed from the digits 0, 1, 2, 3 and 4 if the digits cannot repeated in the same numeral?
    (ii) How many numbers in c(i) above are odd?
    (iii) Find the probability that the number formed in c(i) above is greater than 2000.

Examination Paper 7

TIME: 3 hours

Instructions

1. This paper consists of two sections A and B.
2. Answer **all** questions in section A and **any four** questions in section B.
3. All work done must be shown clearly.
4. Don't use electronic calculators for any question.
5. You are advised to spend no more than 2 hours in section A and using the remaining time in section B.

Section A

1. (a) If $p = 0.18$ and $q = 0.27$, write the value of
   (i) pq
   (ii) $\dfrac{q}{p}$ in the form of $a \times 10^n$ where $0 \leq a < 10$ and n is any integer.

   (b) Without using mathematical tables, evaluate
   $$\left(\frac{8}{27}\right)^{-\frac{2}{3}} + \frac{(64 \times 125)^{\frac{1}{3}}}{\sqrt[4]{81^2}}$$

2. (a) If $p^{5+2a} \times q^{4a+2} = p^{4+a} \times q^{5a+2}$, show that
   $$\log p = a \log \frac{q}{p}$$

   (b) Given that $\log 2 = 0.3010$ and $\log 3 = 0.4771$. Find

   (i) $\log 0.36$ (ii) $\log 6\dfrac{2}{3} + \log 2\dfrac{1}{2}$

   (c) Simplify the following:
   $$\frac{5 - \sqrt{2}}{\sqrt{2} + 4} + \frac{\sqrt{2} + 5}{4 - \sqrt{2}}$$

3. (a) A universal set µ = {Natural numbers less than 10}, and its subsets are A = {Prime numbers}, B = {Odd numbers} and C = {Multiples of 3}. Represent the given sets on Venn diagram, then find
   (i) (A∩B)
   (ii) (A∪B)∩(B∪C)
   (iii)(A∪B∪C)

   (b) Given that set $A = \{(x,y): 5x + y = -8\}$ and set $B = \{(x,y): 2y - 3x = 10\}$. Find the set A∩B.

4. (a) The cost of books consist of a fixed charge of shs.c and the variable charge which depends on the number of books. The cost of 8 books is shs.27,200 and that of 12 books is shs.40,000. Find the cost of 20 books.

   (b) The variable a varies partly as b^2 and partly inversely as b. Given that $a = 5$ when $b = 1$ and $a = 19$ when $b = 3$. Find
   (i) the equation connecting a and b.
   (ii) the value of a when $b = 6$.

5. (a) Find the resultant force of forces 10 N, 12 N and 15 N acting at a point O.

   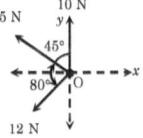

   Figure 3.7.5.

   (b) Solve the system of the following simultaneous equations:
   $$\begin{cases} 2x + 3y - \frac{1}{2}z = 11 \\ x + 3z = 6y - 4 \\ x = z - 2 \end{cases}$$

6. (a) Find the value of y on the figure below.

   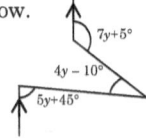

   Figure 3.7.6.

   (b) Calculate the area and perimeter of a regular pentagon inscribed in a circle whose area is 38.5 cm². ($\pi = \frac{22}{7}$)

7. Given that points A(0,*a*), B(*a*-2,7) and C(-2,-2*a*+1) are collinear. Find
   (a) the value(s) of *a*,
   (b) the possible gradient(s) of a line AB, and
   (c) the possible equation(s) of a line AB.

8. (a) The first, third and ninth terms of an arithmetic progression forms the first three terms of a geometric progression. If the sum of the first four terms of the geometric progression is 280, find
   (i) the common ratio and the first term of the geometric progression.
   (ii) the common difference and the first term of the arithmetic progression.
   (b) Fatma has a deposit of $195,000 in her non-interest bank account. In the first month, she drew $2,500. The amount she is drawing increases by $1,500 each month.
   (i) How much in total would she have drawn after one year?
   (ii) After how long will she have nothing in her account?

9. (a) The figure below is a circle PQRS with ∠QRS = 80°, ∠PQS = 20°, and O as the centre.

Figure 3.7.9(a).

Calculate ∠POQ.

(b) The figure 3.7.9(b) below is a sector of radius 11.5 cm. Calculate
   (i) the perimeter of a sector, and
   (ii) the area of a sector.

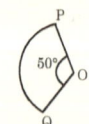

Figure 3.7.9(b).

10. (a) Evaluate the following without using mathematical tables.
    (i) sin75°
    (ii) tan165°
   (b) If $\sin\theta = \frac{1}{2}$ and $\cos\alpha = \frac{1}{3}$, where θ and α are acute angles, evaluate
    (i) $\sin(\theta - \alpha)$ (ii) $\cos(\theta + \alpha)$

Section B

11. A stationeries' seller wants to buy some boxes of dozens of pencils, and of pens for his business. The costs of each pencil and each pen (in full prices) are shs.100 and shs.200 respectively. He wants to spend no more than shs.60,000 on the stationeries. He also wants to buy at least 20 boxes of pencils and 10 boxes of pens. Then, he would retail each pencil and pen for shs.160 and shs.300 respectively.
    (a) How many boxes of each item should he buy so as to maximize his profit?
    (b) What is his maximum percentage profit?
    (c) How many boxes of each item will give the minimum profit?
    (d) What is the difference of the profits obtained in (a) and (c) above?

12. The heights of 100 pupils in a certain school were recorded as follows.

Height (cm)	100–105	105–110	110–115	115–120	120–125
Frequency	16	23	30	20	11

   (a) Calculate the mean height of each pupil.
   (b) Draw a cumulative frequency curve and use it to estimate
    (i) the median height.
    (ii) the number of pupils who are between 107 cm and 113 cm tall.

13. The volume of the right pyramid VPQRS whose base is square PQRS and height is 8 cm is 216 cm³. Calculate
    (a) the length of the slanting edges,
    (b) the angle between the plane VPQ and the base, and
    (c) the surface area of the pyramid.

14. A dealer commenced a business on 1st August, 2005 with Capital in cash $80,000.

   August 2, bought goods for cash $38,000
   3, bought furniture for cash $25,000
   5, paid carriage $800
   9, cash sales $45,000
   13, sold goods to Ali on credit $15,000
   16, purchased goods for cash $25,500
   20, sold goods for cash $49,500
   22, Ali paid a dealer $10,000
   26, cash sales $25,400
   28, paid wages $2,000

   (a) Open the following accounts and balance them at the end.
   - (i) Cash account
   - (ii) Purchases account
   - (iii) Sales account
   - (iv) Ali's account

   (b) Extract a trial balance.

15. (a) A square has its vertices at O(0,0), A(3,0), B(3,3) and C(0,3).
   (i) Draw the square and its image on the same axis under the transformation whose matrix is $\begin{pmatrix} 2 & 0 \\ 0 & 2 \end{pmatrix}$.
   (ii) Find the scale factor area of the transformation.

   (b) A certain linear transformation T is such that T($\underline{u}$) = (2,5) and T($\underline{v}$) = (0,-4) for any vectors $\underline{u}$ and $\underline{v}$. Find
   - (i) T(5$\underline{u}$)
   - (ii) T(2$\underline{v}$ − $\underline{u}$)
   - (iii) T if $\underline{u}$ = (5,3) and $\underline{v}$ = (-4,-3)
   - (iv) T(-7,4)

16. (a) Draw the graph of $f(x) = 2 - |x|$

   (b) The probability that John goes to school earlier before the time is $\frac{1}{8}$ and the probability that he goes to school late is $\frac{1}{4}$. If he goes to school before the time, he doesn't forget anything at home. If he goes to school on time, the probability that he forgets something at home is $\frac{1}{10}$ and if he goes to school late, the probability that he forgets something is $\frac{2}{5}$. Draw a tree diagram to represent the information above, and hence find the probability that John
   (i) forgets something at home when he goes to school.
   (ii) doesn't forget anything when he doesn't go on time.

Examination Paper 8

TIME: 3 hours

Instructions

1. This paper consists of two sections A and B.
2. Answer **all** questions in section A and **any four** questions in section B.
3. All work must be done must be shown clearly.
4. Don't use electronic calculators for any question.
5. You are advised to spend no more than 2 hours in section A and using the remaining time in section B.

Section A

1. (a) A car was sold on shs.6,698,750. This price includes 15% of VAT. Calculate the amount of VAT.
   (b) (i) Write 0.008358 in the standard notation correct to three significant figures.
   (ii) If $x*y$ is defined as $(x+y)^2 - y^2$, find $(1*3)*(-1*2)$.

2. (a) Solve for a and b from the following simultaneous equations:

   $$\sqrt{8\log_{5b} a} = \log_a 5b \text{ and } \log(a+b) = 1$$

   (b) Solve for the inequality $2 < |x+6| \leq 8$.

3. (a) In a school of 320 students, there are 140 boys and 101 students who do not play games. The number of girls who play games is 95. How many boys play games?
   (b) Given that the number of elements of a universal set μ is 40, n(A) = 20, n(B) = 23, and n(A∪B) = 35, where A and B are subsets of μ. Find n(A∩B') and n(A'∪B).

4. (a) Given the points A(2,3), B(-2,5) and C(4,-3).
   (i) Find the vectors $\overrightarrow{AB}, \overrightarrow{AC}$ and $\overrightarrow{CB}$.
   (ii) Verify that $\overrightarrow{AB} + \overrightarrow{CA} = \overrightarrow{CB}$.
   (iii) Calculate the magnitude of $\overrightarrow{CB}$.

   (b) A ship set out 200 km/h on a course of 120° and met a north easterly current of 80 km/h. Calculate the ship's
   (i) ground speed,
   (ii) drift, and
   (iii) track.

5. The surface area of a cylinder, A is partly varying as the square of its radius r, and is partly varying jointly as the radius r and its height h. When $r = 3.5$ and $h = 8$, $A = 253$ and when $r = 10.5$ and $h = 4$, $A = 957$. If the constants of proportionality are a and b respectively,
   (a) formulate A in terms of r and h. (Leave the constants in fraction form)
   (b) find A when $r = 7$ and $h = 10$.
   (c) If $a = x\pi$ and $b = y\pi$ where $\pi = \dfrac{22}{7}$, find x and y, hence rewrite the formula expressing A in terms of π, r and h.

6. Draw the graph of $y = x^2 + 2x - 8$ for the values of x, $-5 \le x \le 4$ and then use it to solve the following equations:
   (a) $x^2 + 2x - 8 = 0$
   (b) $x^2 - 2x - 3 = 0$

7. (a) On the picture of a certain village, the area of the plan of a certain building is 25 cm², while the actual area of the plan of the building is found to be 0.0625 km². Find
      (i) the scale of that picture in one dimension, and
      (ii) the volume of the tank on the actual area in m³ if it measures 0.15 cm by 0.08 cm by 0.2 cm on the picture.
   (b) The areas of two similar triangles are in the ratio 5:8. Calculate the area of the larger triangle if the area of the small one is 10 cm².

8. (a) The point X divides the line segment PQ internally in the ratio 1:2. Given that the coordinates of P and Q are (2,-5) and (6,3) respectively. Find
      (i) the coordinates of X.
      (ii) the distance $\overline{QX}$.
   (b) The equation of a circle is $x^2 + y^2 - 6x + 10y = 15$. Write the equation in the form of $(x-a)^2 + (y-b)^2 = r^2$, hence find
      (i) the centre and radius of the circle
      (ii) the area of the circle.

9. (a) The sum and product of the first three terms of a geometric progression are 37 and 1728 respectively. Find
   (i) the common ratio,
   (ii) the three numbers, and
   (iii) the sum of terms of the progression to infinity when the common ratio is less than one.
   (b) The sum of the first 20 terms of an arithmetic progression is 920 and the sum of the next 10 terms is 1060. Find the sum of the first ten terms and the general formula of the progression.

10. (a) The area of a regular decagon which is inscribed in a circle is 1175.6 cm². Calculate the area and circumference of the circle.
    (b) The circles of diameters 12 cm and 4 cm (on figure 3.8.10) have their centres 17 cm apart. Find the length of the common tangent line $\overline{AB}$ at points A and B of the circles.

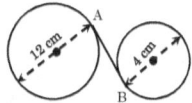

Figure 3.8.10.

Section B

11. (a) Given that $f(x) = ax^2 + bx + c$ such that $f(1) = 7$, $f(2) = 14$ and $f(3) = 25$. Find a, b and c.
    (b) A dealer has two packaging deposits, A and B holding 100 and 80 packets of rice respectively. He has two agents, C and D, who have ordered 70 and 50 packets of rice respectively. If x and y packets of rice are to be sent to agents C and D respectively from the deposit A, formulate six inequalities from the information and use them to draw their graph and show the feasible region.

12. The weights of 50 people are recorded in the table below.

Weight (W) in kg	60 ≤ W < 70	70 ≤ W < 80	80 ≤ W < 90	90 ≤ W < 100
Frequency	10	17	14	9

(a) Draw a histogram and use it to estimate the mode weight.
(b) Draw a cumulative frequency curve and use it to estimate
   (i) the median weight.
   (ii) the number of people of weight above 75.

13. The volume of a geographical globe is 1000 cm³.
    (a) Calculate its radius, and hence find its surface area.
    (b) Calculate the distance between two points A(25°N, 30°E) and B(35°S, 30°E) on the geographical globe.

14. From the following trial balance of PRT Company, prepare a trading, profit and loss accounts for the month ended 29th February, 2008 and a balance sheet as at that date.

Account name	Dr ($)	Cr ($)
Capital	-	5,000,000
Purchases	6,560,000	-
Sales	-	9,900,000
Cash	1,500,000	-
Return outwards	-	584,700
Return inwards	675,820	-
Opening stock	978,780	-
Salaries	778,000	-
Wages	685,000	-
Rent	136,300	-
Insurance	1,200,000	-
Fixture and fittings	800,000	-
Motor vehicles	750,000	-
Debtors	820,800	-
Creditors	-	400,000
Drawings	1,000,000	-
Total	15,884,700	15,884,700

Note closing stock in hand was $860,000

15. (a) Given that matrix $P = \begin{pmatrix} 0 & 1 \\ -2 & 3 \end{pmatrix}$, $Q = \begin{pmatrix} 4 & 2 \\ 1 & -2 \end{pmatrix}$ and $R = \begin{pmatrix} 0 & 2 \\ 3 & 1 \end{pmatrix}$

Find
    (i) the matrix $(P+Q)^{-1}$
    (ii) the matrix X if $PQ+X^{-1} = R$.

(b) Find the area of the image obtained after the plane of area 2.5 square units transformed by the transformation of matrix $\begin{pmatrix} 2 & 5 \\ -3 & 4 \end{pmatrix}$.

(c) Enlarge the triangle PQR with vertices P(-2,3), Q(1,4) and R(3,2) by the enlargement factor 2 about the point (1,3).

16. (a) Given that $f(x) = \begin{cases} 2 & \text{when } -3 < x < 1 \\ x^2 - 1 & \text{when } x \geq 1 \end{cases}$

   (i) Sketch the graph of f.
   (ii) State the domain and range.
   (iii) Find $f(-2)$ and $f(-6)$.

(b) A box contains four blue balls, six red balls and five yellow balls. If one ball is drawn at random without a replacement and then another ball drawn at random, find the probability that
   (i) the first ball is blue and the second is yellow.
   (ii) the first ball is red and the second is either blue or yellow.
   (iii) the two balls are both yellow.

Examination Paper 9

TIME: 3 hours

Instructions

1. This paper consists of two sections A and B.
2. Answer **all** questions in section A and **any four** questions in section B.
3. All work done must be shown clearly.
4. Don't use electronic calculators for any question.
5. You are advised to spend no more than 2 hours in section A and using the remaining time in section B.

Section A

1. (a) Use mathematical tables to evaluate $\left(\sqrt{25.8} + 3.24^2\right)^3$

   (b) Write $0.80\dot{3}$ in the form of $\dfrac{a}{b}$ where a and b are integers, and $b \neq 0$.

2. (a) Find the value of x and y if $\dfrac{(3^{2x})(5^{6y})}{(25^y)(3^{4x})} = 15^2$

   (c) Given that $\log 2 = 0.3010$, without using mathematical tables, evaluate
   (i) $\log 50$
   (ii) $\log 0.25$

3. (a) μ is a universal set such that, $n(\mu) = 18$. A and B are the subsets of μ where $n(A) = 11$, $n(A \cap B) = 5$ and $n(B) = 8$. Find
   (i) $n(A')$ (ii) $n(A \cup B)'$

   (b) Without using any mathematical aids, evaluate
   $$\dfrac{\sqrt{2}+9}{\sqrt{2}-1},$$
   given that $\sqrt{2} = 1.414$

4. (a) Given that vector **a** = 3*i* − *xj*, 3**b** = 3*xi*+14*j* and **c** = 2**a**+3**b**. Calculate the value of *x* if |**c**| = 17.
   (b) On the ifgure 3.9.4 below, ABCE is a cyclic quadrilateral, $\overline{BD}$ is a diameter of the circle and $B\hat{C}E$ = 80°. Calculate
   (i) $B\hat{A}E$ (ii) $E\hat{B}D$

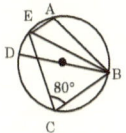

Figure 3.9.4.

5. (a) Find the angle *x* from the figure below.

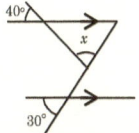

Figure 3.9.5.

   (b) Use the idea that $(a - 1) \times b = ab - b$ to evaluate 999×55.
   (c) Factorize completely the expression $x^2y+3xy+2y+x+2$

6. The volume of a sphere, *V*, is directly proportional to the cube of its radius, *r*. The volume is 38808 cm³ when the radius is 21 cm.
   (a) Find *V* when *r* = 10 cm.
   (b) Write the formula for *V* in terms of *r*.
   (c) If the constant of proportionality from the formula is equal to $a\pi$ where $\pi = \frac{22}{7}$, find the value of *a*, hence write the formula containing *V*, *r* and π.

7. The line segment PC is the perpendicular bisector of the line segment AB at X. Points A and B are (4,6) and (-2,4) respectively and the equation of the line BC is $x+2y = 6$. Find
   (a) the point C.
   (b) the area of triangle BCX.

8. (a) Find the sum of the series $1+\frac{1}{3}+\frac{1}{9}...$ to infinity.
   (b) The sum of the second term and the term before the last term of an arithmetic progression is 50 and the sum of third and fourth terms is 34. If the last term is 20 greater than the fifth term, find the first term and the common difference.

9. (a) Three brothers, Ali, Juma and Soud shared some amount money. The ratio of Ali's that of Juma's share is 3:2 and the ratio of Juma's to Soud's share is 3:4. If Soud got $1,600, find the shares of Ali and Juma.
   (b) The length of the chord of a large circle is 10 cm and the lenght of the corresponding chord of a small circle is 7 cm. Find the area of the small circle if the area of the large circle is 400 cm².

10. (a) The angles of elevations from the bottom and the top of a cliff to the building 20 m apart are 60° and 30° respectively. Find the height of the building and the cliff.
    (b) Solve the following triangle.

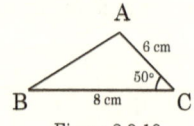

Figure 3.9.10.

Section B

11. A small factory makes two types of products A and B. Each of type A product takes 7 hours and $50 worth of chemicals to make and each type B takes 4 hours and $60 worth of chemicals to make. The factory has $620 for chemicals, to make the products in 56 hours. The profit they make on type A is $28 shillings and the profit on type B is $30 per product. How many of each type of products should the factory make in order to make the profit as maximum as possible?

12. The following frequency distribution table shows the scores of 100 students in a geography exam.

Scores	0–20	20–40	40–60	60–80	80–100
Frequency	19	20	35	16	10

(a) Draw a histogram to represent the information above and use it to estimate the mode.
(b) Use an assumed mean as 50, to calculate the actual mean.
(c) Use an appropriate formula to calculate the median.

13. (a) From the following figure, $B\hat{X}C = 30°, B\hat{O}C = 20°$, and the radius of the circle is 10 cm. O is the centre of the circle. Find the area of the shaded region.

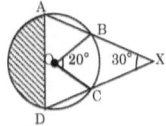

Figure 3.9.13.

(b) Towns A and B are at (25°N, 36°E) and (25°N, 40°W) respectively. Find
   (i) the time at B when at A is 1:50 a.m.
   (ii) the distance from town A to B.

14. The following trial balance was extracted from the books of PQR Company.

Account	Dr ($)	Cr ($)
Capital	-	xxx
Cash	72,000	-
Opening stock	50,000	-
Purchases	xxx	-
Sales	-	xxx
Wages	24,000	-
Rent	xxx	-
Rates	6,000	-
Total	323,580	323,580

Given that

the cost of goods available for sale: $211,580,
the cost of goods sold: $159,580,
and the gross profit: $64,000

(a) Complete the trial balance above.
(b) Determine the closing stock and net profit.
(c) Prepare the balance sheet as on 31st October, 2002.

15. (a) The inverse of a matrix A is $\begin{pmatrix} 2 & 3 \\ 4 & -5 \end{pmatrix}$. Find the matrix A.

(b) Use the result obtained in (a) above to solve the simultaneous equations
$$\begin{cases} 2x + 3y = -1 \\ 4x - 5y = 9 \end{cases}$$

(c) The point P(-2, 4) reflected along the line $y = -x$ and then rotated by 180° clockwise direction about the origin. Find
   (i) The single transformation T of the double transformation and state what kind of transformation does T represent?
   (ii) The image of point P under transformation T.

16. (a) The function $f(x) = 8 - 2x^2 - 4x$. Write f in the form of $f(x) = a(x+b)^2 + c$, hence determine
   (i) the turning point,
   (ii) the axis of symmetry,
   (iii) the maximum or minimum value, and
   (iv) the range of f.

(b) Given that the probabilities $P(A) = \frac{1}{4}$, $P(A \cup B) = \frac{1}{2}$ and $P(B') = \frac{2}{3}$ where A and B are not mutually exclusive events. Find $P(A \cap B)'$.

(c) The probability that Asha will be on the first position in the class is $\frac{2}{5}$ and the probability that Juma will be on the first position is $\frac{1}{3}$. If only one pupil will be on the first position in the class, find the probability that
   (i) Asha or Juma will be on the first position.
   (ii) What can you say about the events of Asha and that of Juma being on the first position in the class?

Examination Paper 10

TIME: 3 hours

Instructions

1. This paper consists of two sections A and B.
2. Answer **all** questions in section A and **any four** questions in section B.
3. All work done must be shown clearly.
4. Don't use electronic calculators for any question.
5. You are advised to spend no more than 2 hours in section A and using the remaining time in section B.

Section A

1. (a) Given that $a = 22.47$, $b = 52.03$ and $c = 24.99$
      (i) Round off the values of a, b and c into three significant figures.
      (ii) Evaluate $\dfrac{ab}{c}$ using the answer obtained in a(i) above and estimate the answer into one significant figure.
   (b) Show that $0.56\dot{0}\dot{6}$ is a rational number by writing it in the form of $\dfrac{a}{b}$ where a and b are integers and $b \neq 0$.

2. (a) The LCM and GCF of two numbers x and 28 are 420 and 2 respectively. Find the possible value(s) of x.
   (b) Find the value of n if $3(2)^n = 4^n$.
   (c) If $a^2 + b^2 = 2ab$, show that
   $$\log(a+b) = \log 2 + \frac{1}{2}(\log a + \log b)$$

3. (a) Each student in a class must study at least one subject between science, arts and mathematics. Those who study science must study mathematics. Eleven student's study science, sixteen study arts, twenty-two study mathematics, four study all the three subjects, and nine study arts and mathematics.
      (i) Draw a Venn diagram to represent the information above.
      (ii) How many students study arts but not science?
      (iii) How many students are in the class?
   (b) Given that $\sin\theta = \dfrac{4}{5}$ and $\cos\alpha = \dfrac{5}{13}$ where θ and α are acute angles. Without using tables, evaluate
      (i) $\cos(\alpha - \theta)$ (ii) $\sin(\alpha - \theta)$

4. (a) Given that vectors **a** = 2*i* − 3*j*, **b** = 4*j* and **c** = 2**a**+3**b**. Calculate |**c**|.
   (b) Two forces 30 N and 60 N act together on a load at point O as shown on the figure 3.10.4 below. Find the magnitude of the resultant force.

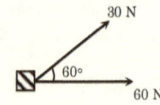

Figure 3.10.4.

   (c) Solve for x and y from the following simultaneous equations:

$$\begin{cases} 2545x + 3435y = 15395 \\ 3435x + 2545y = 14505 \end{cases}$$

5. (a) Factorize completely the expression $(2x+6)^4 - (2x-4)^4$
   (b) Find the value of a, b and c if,

$$12x - 2x^2 - 22 = a(x+b)^2 + c$$

6. The equation of a tangent of a circle whose centre is at point (10,-12) is $15y = 8x + 29$.
   (a) Find the area of the circle.
   (b) Calculate the circumference of the circle.
   Use π = 3.14

7. (a) A variable P partly varies as a variable Q and partly varies inversely as the same variable Q. Given that P = 6 when Q = 1 or Q = 2. Find
   (i) the equation connecting P and Q.
   (ii) P when Q = 4.
   (b) Pipes X and Y can fill a tank in 3 and 4 hours respectively, and pipe Z can empty the tank in 2.5 hours. How long would it take to fill the tank with all three pipes running?

8. (a) The sum of the first four terms of an arithmetic progression whose number of terms is odd number, is 50. The middle term and the sum of the last two terms are 26 and 85 respectively. Find the common difference and the first term.
   (b) The sum of the first two terms and the sum of the first six terms of a geometric progression are 12 and 1092 respectively. Find the common ratio and the first term.

9. (a) A regular hexagon is inscribed in a circle of radius 14 cm. Find the area of the polygon.
   (b) From the circle ABCDE below, ∠EFD = 25° and ∠EAD = 30°. Find ∠BOC if O is the centre of the circle.

   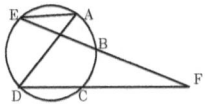

   Figure 3.10.2.

10. (a) 15 cm on the map represents 30 km on the ground. Find
    (i) The scale of the map.
    (ii) The area on the ground in km² if it is found to be 20 cm² on the map.

    (b) From the following triangle, $\overline{BC}$ = 2 cm, $\overline{DE}$ = 6 cm and $\overline{CE}$ = 10 cm. The line BC is parallel to the line DE. Find
    (i) $\overline{AC}$ (ii) $\overline{AE}$

    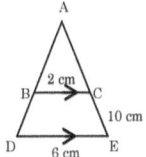

    Figure 3.10.10.

Section B

11. A woman wants to make two types of cakes; type A and type B. To make the cakes, she needs at least 56 eggs. The maximum amount of flour and sugar she needs are 30 kg and 12 kg respectively. Type A cake needs 7 eggs, 1 kg of sugar, and 2 kg of flour to make. Type B cake needs 8 eggs, 1 kg of sugar and 3 kg of flour to make. If she is to make the profits of shs.2,000 and shs.2,500 for type A cake and type B cake respectively,
    (a) Find the maximum and minimum profit.
    (b) How many eggs would she use to make the cakes for a maximum profit?

12. The distribution of ages of 100 people in a certain village is recorded on the table below.

Class mark	5	9	13	17	21
Frequency	9	10+2x	28	22	18-x

   (a) Find the value of x.
   (b) Draw a cumulative frequency curve and use it to estimate the median.
   (c) Find the mode age.

13. (a) The angle between the plane VAB and the base ABCD of area 560 cm² of the right triangular prism below, with height 10 cm is 40°.

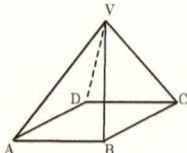

Figure 3.10.13.

   (i) Find the angle between the plane VBC and the base ABCD.
   (ii) Determine the volume of the prism.

   (b) A ship with speed 120 km/h sails from point A(30°N, 106°W) to point B(30°N, 74°E) along the shortest distance.
   (i) How far would it take for the ship to get to the point B?
   (ii) How long would it take to get there?

14. The following information was written from the books of XYZ Company.

   Cost of goods available for sale: $95,000.
   Opening stock: 10% of cost of goods available for sale.
   Closing stock: 5% of sales.
   Gross profit: 10% of sales.
   Expenses: $7,500
   Net profit: 50% of closing stock.

   Prepare a trading, and profit and loss accounts from the information above.

15. (a) Given that matrix $A = \begin{pmatrix} 1 & 2 \\ 3 & 4 \end{pmatrix}$ and $B = \begin{pmatrix} 6 & 9 & 12 \\ 12 & 19 & 26 \end{pmatrix}$.
    (i) Find the inverse of matrix A.
    (ii) Find the matrix X if AX = B, and state the order of matrix X.
    (b) Point P(3,5) is translated by transformation
    $$T = \begin{cases} x' = x + 5 \\ y' = y - 5 \end{cases}$$
    Find the image of point P.
    (c) A linear transformation T maps the position vectors **a** and **b** to **a'** to **b'** respectively. If **a** = $2i+5j$, **b** = $3i+4j$, **a'** = $3i+8j$ and **b'** = $4i+7j$, find T(2**a**+3**b**).

16. (a) A function f is defined as $f(x) = \begin{cases} -x & when\ x > 0 \\ x^2 & when\ x < 0 \end{cases}$
    (i) Sketch the graph of f.
    (ii) State the domain and range of f.
    (iii) Find $f(-3)$.
    (iv) State whether f is one-to-one function.
    (b) A faction is formed from a sets of numerators {1,2,3,4,5} and denominators {7,8,9,10}.
    (i) How many fractions can be formed?
    (ii) What is the probability that a fraction formed is less than $\frac{1}{2}$?

Examination Paper 11

__TIME: 3 hours__

Instructions

1. This paper consists of two sections A and B.
2. Answer **all** questions in section A and **any four** questions in section B.
3. All work done must be shown clearly.
4. Don't use electronic calculators for any question.
5. You are advised to spend no more than 2 hours in section A and using the remaining time in section B.

Section A

1. (a) Approximate each of the following numbers, correct to three significant figures.

    24.38   19.53   52.04

    (b) Use the results obtained in (a) above to show that
    $$\frac{24.38 \times 19.53}{52.04} \approx 9\frac{3}{20}$$

    (c) Find the quotient when the Lowest Common Multiple of numbers 54, 63 and 72 is divided by their Highest Common Factor.

2. (a) Solve for x from the equation
    $$\left(\frac{1}{27}\right)^x \times 9^{2x-1} \times 135 = \left(\frac{81^x}{5}\right)^{-1}$$

    (b) Given that $\log_x a = 0.4$ and $\log_y a = 0.8$. Evaluate t if $y^2 a^t = \sqrt{x^3}$.

3. (a) Given that A and B are the subsets of μ such that A' = {f,g,h,i}, B' = {c,d,e,h,i}, and A∩B = {a,b}. Draw a Venn diagram for sets A, B and μ, hence find
    (i) μ   (ii) A∪B

    (b) Two sets A and B of universal set U are such that n(A) = 8, n(B') = 7, n(A'∪B') = 15, and n(U) = 18. Use a Venn diagram to find
    (i) n(B)   (ii) n(A∪B)

4. (a) Point C(*t*,2*t*) is on the perpendicular bisector of a line joining points A(2,3) and B(-4,1). Find
   (i) the value of *t*.
   (ii) the area of △ABC.
   (b) Given that vectors **a** = 2***i***+3***j*** and **b** = 3***i*** − 2***j***. Find | **a** − **b** |. What are the direction cosines of **a** − **b**?

5. (a) From the right angled triangle below, find the length *h*, perpendicular from A to BC.

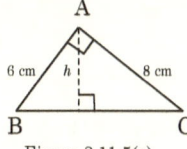

Figure 3.11.5(a).

(b) The triangle below is an isosceles triangle such that AB = AC.

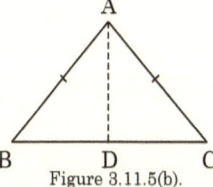

Figure 3.11.5(b).

If AD is a bisector on the line BC, prove that △ADB ≡ △ADC, and hence prove that ∠ADC is a right angle using the fact that BDC is a straight line.

6. (a) Given that *p* is directly proportional to q^2 whereby *p* = 1 when *q* = 2. Find *p* when *q* = 6.
   (b) An old man can dig a trench in 10 days. When assisted by his son, they can do the work together in 6 days. How long would his son take to do the work by himself?

7. (a) John, Peter and Adam shared capitals of £45,000, £67,500, and £90,000 respectively for a business under agreement that the shares of the profit were proportional to the capital provided. If the profit were £18,000, what should each one receive?
   (b) An increase of salaries of 15% makes the monthly salary of a person being £32,200. What is the amount of increase?

8. (a) Find the three numbers in an arithmetic progression such that their sum is 36 and their product is 1428.
   (b) The sum of the fifth and seventh terms of a geometric progression is eight times the sum of the second and fourth terms of the same progression. Find the common ratio of the progression.
   If the sum of the first six terms of this progression is 189, find its first term.

9. (a) The exterior angle of a regular polygon is θ. Prove that the sum of all interior angles of the polygon is given by $\dfrac{360(180° - \theta)}{\theta}$, provided that the sum of all exterior angles is always 360°. Use the formula to find the sum of interior angles of a regular octagon.
   (b) Without using mathematical tables, find the numerical value of $2\tan^2 30° + 6\cos^2 45° + \dfrac{4}{\sin^2 60°}$.

10. (a) Given that $x^2+2x = 3$, prove that $x^5 = 61x - 60$.
    (b) By completing the square, solve the inequality $x^2+2x - 15 \leq 0$.

Section B

11. A certain linear programming problem led the following constraints.

    $$5x+y \geq 15$$
    $$x+y \leq 7$$
    $$3x+2y \leq 18$$
    $$x \geq 0, y \geq 0$$

    The information needed to maximize the constraints subject to the objective function $F(x,y) = 3x+4y$.
    (a) Draw the graph to illustrate the given inequalities.
    (b) Use the corner points to find the maximum value of F.

12. The marks scored by students in a mathematics test were recorded on the following frequency distribution table.

Boundaries	40–50	50–60	60–70	70–80	80–90
Frequency	14	16	8	6	6

    Calculate (a) the mean,
    (b) the mode, and
    (c) the median scores.

13. (a) Locate two places A and B on the same diagram of the earth's surface if A is at (25°N, 38°E) and B is at (65°S, 38°E). Calculate the distance AB on the earth's surface.
    (b) Prove that any two tangents from an external point of the circle are equal.
    (c) Find the radius of the circle below if AB and AD are the tangents at B and D respectively and that ∠BCD = 30°, and AB = 10 cm.

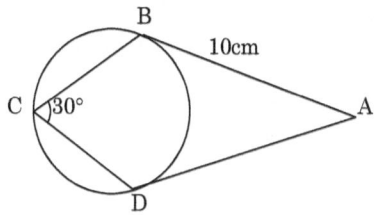

Figure 3.11.13.

14. Record the following transactions in cash, purchases, sales, Makame's, and Asha's accounts.

    On 1st May, Zahor commenced business with capital in cash $80,000
    May 2: bought goods for cash $35,000
        5: sold goods on credit to Makame $28,000
        9: sold goods in cash $30,000
        13: bought goods from Asha on credit $25,000
        16: paid wages in cash $6,000
        21: paid Asha in cash $7,000
        25: paid rent in cash $4,500

    Balance the accounts as on 31st May and extract the trial balance.

15. (a) Given that matrix $A = \begin{pmatrix} 2 & 4 \\ 1 & 3 \end{pmatrix}$ and $B = \begin{pmatrix} -9.5 & 6 \\ 1.5 & -8 \end{pmatrix}$. Find AB+5A, and then deduce A^{-1}.

    (b) P_1 is a point (2,5). The matrix $L = \begin{pmatrix} 2 & 3 \\ 1 & 0 \end{pmatrix}$ maps P_1 onto P_2 and the matrix M maps P_2 onto P_3. If M represents a matrix which maps a point along the line $x+y = 0$, show that the matrix LM maps P_1 onto P_3.

16. (a) Two students, Ali and Juma did an examination. The probability that Ali will pass the exam is $\frac{1}{3}$, and the probability that Juma will pass the examination are and $\frac{3}{5}$ respectively. Find the probability that
   (i) Ali will not pass   (ii) Juma will not pass
   Hence find the probability that at least one of them will pass the examination.
(b) The function $f(x) = 2x^2+4x - 5$. Show that this can be written in the form of $f(x) = a[g(x)]^2+c$ where a and c are real numbers. State the turning point, axis of symmetry and the range of the function. Sketch its curve.

Examination Paper 12

TIME: 3 hours

Instructions

1. This paper consists of two sections A and B.
2. Answer **all** questions in section A and **any four** questions in section B.
3. All work done must be shown clearly.
4. Don't use electronic calculators for any question.
5. You are advised to spend no more than 2 hours in section A and using the remaining time in section B.

Section A

1. (a) Evaluate the following, correct to four significant figures.

    $$\frac{1}{7}\sqrt{\frac{5.25}{2\pi}}$$

    given that $\pi = 3.142$

    (b) The average age of a class of 40 students was 16 years, and 6 months. When one of the students left the class, the average of the remaining students was 16 years, and 7 months. What was the age of the students who left the class?

2. (a) Simplify the following expression:

    $$\frac{\log_a b + \log_{\sqrt{a}} b^2}{\log_a b - \log_{\sqrt{a}} b^2}$$

    (b) Given that $\sqrt{2} = 1.414$ and $\sqrt{3} = 1.732$, without using tables, evaluate $\dfrac{\sqrt{6}}{\sqrt{3}-\sqrt{2}}$.

3. (a) Given that, the universal set $\mu = \{0,1,2,a,b\}$ and its subsets $A = \{0,1,2\}$ and $B = \{2,a\}$. Represent the given sets on Venn diagram and use it to find
    (i) $A - B$ (ii) $(A \cap B)'$

    (b) Make n as the subject of the formula $l = \dfrac{nQ}{R+nP}$.

4. (a) On the diagram below, $\overrightarrow{BC} = \underline{a}$ and $\overrightarrow{AC} = \underline{b}$. P and Q are the midpoints of BC and AC respectively.

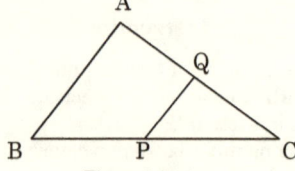

Figure 3.12.4

Find the vectors $\overrightarrow{BC}$ and $\overrightarrow{QP}$. Hence show that $\overrightarrow{AB} = k\overrightarrow{QP}$ State the value of k.

(b) Given that vector $\underline{r} = k\begin{pmatrix} 1 \\ 2 \end{pmatrix}$ where k is a scalar. Find the value of k if $|\underline{r}| = 2\sqrt{5}$.

5. (a) On the figure below, ABCD is a kite; BD is a perpendicular bisector of AC. If AC = a units and BD = b units, find an expression for the area of a kite.

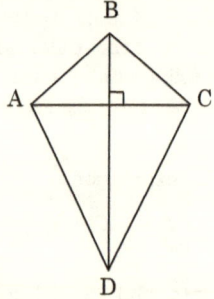

Figure 3.12.5

(b) The scale of a map is such that 1 cm on a map represents the actual length of 2500 cm.
   (i) Find the length on a map whose actual distance is 0.15 km
   (ii) Find the actual area when it is 2.4 cm² on the map.

6. (a) One-third of a piece of work is completed in fourteen days by eighteen men, each working 8 hours a day. How many additional men must be employed in order that the work may be finished in seven more days if each man now working 6 hours a day?
   (b) Given that,
   $$x \propto y^2 \text{ and } y \propto \frac{1}{\sqrt{z}}.$$
   How does y varies jointly with x and z (Use the same notation to represent). Given that $y = 4$ when $x = 4$ and $z = 9$, write the equation connecting y, x and z.

7. (a) Juma bought some coconuts at shs.600 each so as to sell. He sold the rest at shs.3200 per group of 4 coconuts when he found that 20% of coconuts are unsaleable. Find his percentage profit.
   (b) When a certain sum of money was deposited at a bank which gives a rate of 2.5% per annum simple interest for a certain number of years, it amounted to $5,375. If the same principal would have been deposited at a bank which gives a rate of 3.5% per annum simple interest for 2 years more, it would have amounted to $5,875.
      (i) How long did it take for the money at a rate of 2.5% per annum simple interest to mount to $5,375?
      (ii) Find the deposited amount.

8. (a) The sum of the first ten terms of an arithmetic progression is 175 and the sum of the next ten terms of the same progression is 475. Find the sum of the first thirty terms of this progression.
   (b) The sum of the first two terms of a geometric progression is 16. The sum of the first four terms is ten times the sum of the first two terms. If the common ratio is positive, find it and the first terms of the progression.

9. (a) Solve for x from the equation
   $$\tan(x + 60°)\sin x = \cos x$$
   if x is an acute angle.
   (b) The legs of a pair of divider are 5 cm long each. Find the distance between its tips if the legs contain an angle of 80°.

10. (a) A bookseller sells p books at a profit of x% on the buying price and q books at a loss of y%. All the ($p+q$) books cost bookseller the same. Find the expression for his percentage profit on the total sale.

(b) Given that $\dfrac{3x+4}{(x+1)(x+2)} = \dfrac{A}{x+1} + \dfrac{B}{x+2}$, find the values of A and B. Hence, or otherwise, find $\dfrac{1}{6} + \dfrac{1}{7}$.

Section B

11. Kilimanjaro, the boat can take a maximum of 500 passengers. A profit of tzs.30,000 is made on each first class ticket and a profit of tzs.25,000 is made on each economy class ticket. The boat's company reserves at least 100 seats for first class. However, at least 3 times as many passengers prefer to travel by economy class than first class. How many tickets of each type must be sold in order to maximize the profit for the company?

12. The ages in years of the students in the class are given as follows:

    12 11 13 11 10 13 12 15 13 14
    10 12 12 13 14 13 15 11 11 12

    (a) Construct the frequency distribution table for the data and find the mode and median age.
    (b) Calculate the mean age of the class.
    (c) After three years, the age of these students in a class were recorded again. Use the mean obtained in (b) above to find the new mean age of the class after these three years.

13. (a) A small closed box is made of wood everywhere 2 cm thick and its internal dimensions are 14 cm by 10 cm by 12 cm. Find the volume of wood used to make box.
    (b) Find the area of the surface of the box in (a) above.

14. From the information given below, determine the gross profit and the net profit or loss.

Sales	£50,000
Purchases	£35,000
Opening stock	£6,000
Expenses	£25,000
Return outward	£3,500
Return inward	£1,000
Closing stock	£5,800

15. (a) Find the values of x and y if, $3\begin{pmatrix} 1 & 2 \\ 5 & x \end{pmatrix} + \begin{pmatrix} y & 1 \\ 0 & 5 \end{pmatrix} = \begin{pmatrix} 5 & 7 \\ 15 & 14 \end{pmatrix}$

 (b) When the line is rotated by an angle θ about the origin, one of the point remains fixed. The line $2x+y = 5$ is rotated by 90° about the origin, find the fixed point.

 (c) Any point $P(x,y)$ reflects along the line $y = x\tan\theta$ to the point $P'(x',y')$ according to the equation

$$\begin{pmatrix} x' \\ y' \end{pmatrix} = \begin{pmatrix} \cos 2\theta & \sin 2\theta \\ \sin 2\theta & -\cos 2\theta \end{pmatrix} \begin{pmatrix} x \\ y \end{pmatrix}$$

 Find the image of point (2,-4) when reflected along the line $y = \sqrt{3}x$.

16. A farmer has 500 m of a fencing material and he wants to use it to make a rectangular enclosure whose area and one of the sides are A m² and l m respectively.
 (a) Write A as a function of l.
 (b) Write A in the form of $A = a(x+b)^2+c$ where a, b and c are real numbers.
 (c) State the value of l so that the farmer will get the maximum possible area. What is this area?

Examination Paper 13

TIME: 3 hours

Instructions

6. This paper consists of two sections A and B.
7. Answer **all** questions in section A and **any four** questions in section B.
8. All work done must be shown clearly.
9. Don't use electronic calculators for any question.
10. You are advised to spend no more than 2 hours in section A and using the remaining time in section B.

Section A

1. (a) Given that $x = \dfrac{\sqrt{ab}}{c^2}$ such that $a = 10.5$, $b = 13.5$ and $c = 3.4$. Use logarithms tables to evaluate x correct to one decimal place.

    (b) A person used $\dfrac{1}{3}$ of his pocket money on food, and $\dfrac{3}{7}$ on clothes.
    (i) What fraction of the money left?
    (ii) Find the least possible amount of money in dollars that he had before.

2. (a) Solve for x from the logarithmic equation below.
    $$(\log_2 a)(\log_a 8) = \log_x 27$$

    (b) Given that $a = 1 + \sqrt{8}$ and $b = 2 - \sqrt{18}$, simplify
    $$\dfrac{a+b}{a-b}$$
    and rationalize the denominator your answer.

3. (a) In a certain school, students who are taking advanced level studies, 35% take history, 30% take geography and 5% take advanced mathematics. All those who are taking advanced mathematics are also taking geography but not history. If 45% do not take any of these three subjects, what is the percentage of students who are taking
    (i) geography and history?
    (ii) geography only?

    (b) Expand $(1+2x)^3$. Hence use the substitution $x = 0.2$ to find the value of 14^3.

4. The angle that vector **r** makes with the x-axis is α and that makes with y-axis is β where **r** is in the first quadrant of x and y plane.
   (a) Show that **r** = |**r**| (cosα***i***+cosβ***j***)
   (b) If **r** = 5***i***+12***j***, find the value of cosα and cosβ. What do these values represent?
   (c) Deduce the unit vector in the direction of vector **r**.

5. (a) Give reasons why the triangles below are similar? Find the values of the unknown sides and angles.

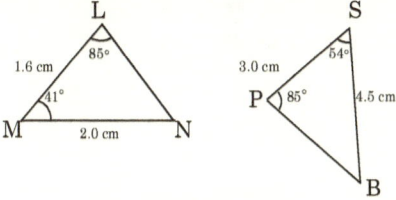

Figure 3.13.5(a).

(b) From the figure below, find the value of x and y.

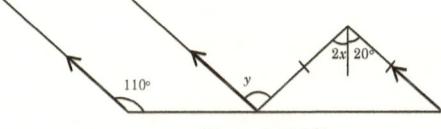

Figure 3.13.5(b).

6. (a) A person can run a distance of 1km in 2 hours, and 20 minutes. How far can the person run in 10 hours, and 30 minutes?
   (b) The product of two numbers x and y is always constant.
      (i) How does y varies with x?
      (ii) Given that $y = 10$ when $x = 2.5$, find the equation connecting y and x.
      (iii) Draw a sketch of the graph of y against x.

7. (a) A wire was divided into three pieces in the ratio 2:4:5. The largest piece was then divided in the ratio 1:3. Find the ratio of the lengths of four pieces of the wire in ascending order (i.e. from the smallest to the largest).
   (b) John bought 10 kg of rice for shs.1,200 per kilogram and 6 kg of sugar for shs.1,500 per kilogram. He paid shs.16,800 after receiving a discount. What was the percentage of discount?

8. (a) The sum of all k terms of a finite arithmetic series of the last term 60, is $40k$. Find the first term of the series if the common difference is 2. How many terms does this series have?
   (b) Calculate the arithmetic mean of numbers 1, 2, 3, 4, 5, 6, 7, 8, 9 and 10. Hence show that the geometric mean of numbers 4, 4^2, 4^3, 4^4, 4^5, 4^6, 4^7, 4^8, 4^9 and 4^{10} is 2^{11}.

9. (a) Prove that $\sin(270° + x) = \cos(180 + x)$. Also verify this identity by substituting $x = 30°$.
   (b) From the triangle ABC below, AC = 18 cm, AB = 15 cm, and BC = 9 cm. Find DC and BD. Also find ∠DBC.

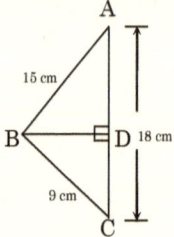

Figure 3.13.9.

10. (a) Using substitution $y = mx$ to solve the simultaneous equations

    $x^2 + 3xy + y^2 = 31$ and $4x^2 = 3xy - 2$

   (b) Solve the inequality $\left|\dfrac{2x-1}{x+1}\right| \geq 1$.

Section B

11. If Eddy rides his motor cycle at 30 km/h, he has to spend shs.5,000 per kilometre on petrol. If he rides at a speed of 40 km/h, the cost increases by shs.2,500 per kilometre. He has shs.240,000 to spend on petrol and wishes to find what is the maximum distance he can travel within one hour. What should he do to fulfil the condition?

12. The table below is the frequency distribution table constructed from a mathematics test scores of 100 students.

Scores	0–20	20–40	40–60	60–80	80–100
Frequency	12	25	38	23	2

(a) Draw a cumulative frequency curve to represent the data and use it to estimate the median.
(b) How many students would fail if the pass mark is 35?
(c) If the top 5% of the students are to be given a grade A, what is the lowest mark which will achieve this?

13. (a) The circle ABCD shown below has O as its centre. Given that ∠AOB = 100°, and ∠AED = 60°. Find ∠COD.

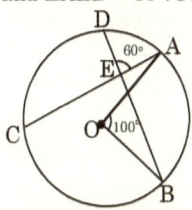

Figure 3.13.13.

(b) A right pyramid has a rectangular base of sides 4 cm and 6 cm and the slanting edges of lengths 10 cm. calculate the height and volume of the pyramid.

14. The following information was obtained from the trading, and the profit and loss accounts.

    Sales: $50,000
    Purchases: $33,000
    Net profit: 24% of sales
    Expenses: 25% of gross profit
    Cost of goods available for sale: 70% of sales.

Determine the following:
(a) Net profit
(b) Gross profit
(c) Expenses
(d) Closing stock
(e) Cost of sales
(f) Opening stock

15. (a) Given that matrix $A = \begin{pmatrix} 1 & 2 \\ 2 & 3 \end{pmatrix}$ and $B = \begin{pmatrix} -2 & 1 \\ 0 & 3 \end{pmatrix}$.
    (i) Find the matrix AB.
    (ii) Verify that $|AB| = |A| \times |B|$.

    (b) The matrix M is $\begin{pmatrix} 6 & 3 \\ 7 & 5 \end{pmatrix}$.
    (i) Find M^{-1}.
    (ii) Verify that $|M^{-1}| \times |M| = 1$.

16. (a) The function f is defined as $f(x) = x^2 - 2x - 4$, x being positive.
    (i) For what domain is f a one-to-one function?
    (ii) Find the inverse of f for the domain obtained above.

    (b) Draw the graph of the inverse of f obtained in a(ii) above. Does this graph represent a function?

Examination Paper 14

TIME: 3 hours

Instructions

1. This paper consists of two sections A and B.
2. Answer **all** questions in section A and **any four** questions in section B.
3. All work done must be shown clearly.
4. Don't use electronic calculators for any question.
5. You are advised to spend no more than 2 hours in section A and using the remaining time in section B.

Section A

1. (a) Three-quarters of a cake is to be divided so that each person gets one-third of it. What is the fraction of the cake (as a whole) will each one get? If the weight of the whole cake is 2 kg, what is the weight of the piece of the cake would each person get?

   (b) Find the value of $(0.85)^{12}$ using logarithmic tables, and approximate the answer correct to 3 decimal places.

2. (a) If a and b are non-zero numbers such that $a = b$, show that $\log(a+b) = \log 2 + \left(\dfrac{\log a + \log b}{2}\right)$.

   (b) Given that $2^x = 1+a$. Show that if $6^x - 3^x = 3^3$, then $x = 3 - \log_3 a$.

3. (a) By the use of Venn diagram for any two sets A and B, prove that $A \cup (A \cap B) = A$.

   (b) Given that, $A' \cup B' = \{1,2,5,6\}$, $A = \{1,2,3,4\}$, and $A' \cap B' = \{5\}$. Find
   (i) the universal set, μ (ii) A' (ii) B'

4. (a) On the figure given below, $\vec{AB} = b$, $\vec{AC} = a$. D divides the line AC such that AD:AC = 2:3.

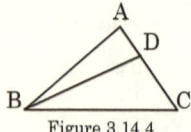

Figure 3.14.4.

Find the vectors $\vec{BC}$ and $\vec{BD}$. Hence find the numerical values of scalars p and q if $\vec{BC} = p\vec{AB} + q\vec{BD}$.

(b) Two forces 12 N and F N, which are inclined at an angle of 120°, have a resultant of magnitude $F\dfrac{\sqrt{21}}{5}$ N. Calculate the value of F.

5. (a) Calculate the area of a parallelogram below if its perimeter is 40 cm.

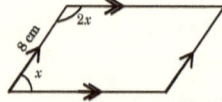

Figure 3.14.5.

(b) Calculate the minimum possible area of a regular pentagon which can be circumscribed by a circle of radius 10 cm.

6. (a) Ali changed $12 into Tanzania shillings (tzs) and he got tzs.14,640. Given that 1 Indian RPS = tzs30. How much would he get if he changed $12 into Indian RPS?
(b) The height by which a certain tree grows is directly proportional to the square of the time taken for it to grow. Given that the tree is 10 m high in 2 years, after how long would the tree be 50 m high?

7. (a) Three people shared £14,000 so that the second got twice as much as the first and the third got twice as much as the second. Find the share of each one.
(b) Fatma marks her goods to gain 40%. She allows 15% discount for cash. Find her percentage profit when she sells for cash.

8. (a) If the thirty-first term of an arithmetic progression is eight times the third term, prove that the thirty-fifth term is six times the fifth term.

(b) Show that the series $1 + \dfrac{1}{x^2} + \dfrac{1}{x^4} + \ldots$, $x \neq 0$ is a geometric series. State the range of values of x for which the sum to infinity exists, and find it's sum to infinity. Hence find the sum of the series $\dfrac{1}{2} + \dfrac{1}{4} + \dfrac{1}{8} + \ldots$ to infinity

9. (a) Prove that $\dfrac{\sin x}{1 - \cos x} = \dfrac{1 + \cos x}{\sin x}$.

(b) From the triangle below, state the values of $\sin\theta$ and $\cos\theta$ in terms of a, b and c.

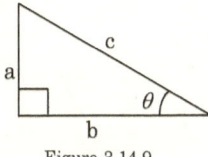

Figure 3.14.9.

Hence use the identity $\sin^2\theta + \cos^2\theta = 1$ to derive the Pythagoras theorem for the given triangle.

10. (a) One regular pentagon has 2 times as many sides as the other. An interior angle of the second polygon is three-quarters of an interior angle of the first. What are the polygons?

(b) Solve the following equations simultaneously.

$$\begin{cases} xy + y = 9 \\ 3y - xy = 3 \end{cases}$$

Section B

11. Two types of fertilizers F_1 and F_2 is such that F_1 consists 10% of nitrogen and 5% of phosphoric acid and F_2 consists of 7% of nitrogen and 9% of phosphoric acid. Experiments show that a farmer needs at least 13 kg of nitrogen and 12 kg of phosphoric acid for her crops. The cost of F_1 is $600 per kg and the cost of F_2 is $700 per kg. How much of each fertilizer should be bought so that the cost is minimum but the fertilizer still meet the requirements?

12. The frequency distribution table below represents the marks scored by students in a certain test and their respective cumulative frequencies.

Class-interval	1–10	11–20	21–30	31–40	41–50
Cumulative frequency	7	25	39	47	50

(a) Calculate the median mark.

(b) Use an assumed mean as the mid mark of the modal class to estimate the actual mean mark of the students.

13. (a) Calculate the shortest distance between the points (85°N, 130°E) and (60°N, 50°W) along the great circles.

(b) From the circle ABCD given below, ∠BAF = 80°, ∠ABD = 60° and ∠BDC = 20°. Find ∠CAD.

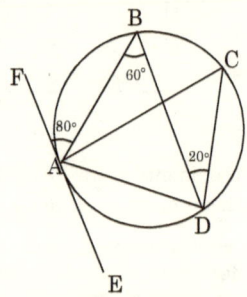

Figure 314.13.

14. The transactions from PQR traders were recorded in cash account as shown below.

(Note: All transactions were paid in cash)

Dr			Cr	
Sept,1 Capital	$4,500	Sept,3 Purchases	$1,500
5 Sales	$2,500	4 Wages	$350
11 Sales	$1,800	13 Insurance	$110
		15 Purchases	$1,150
		20 Drawings	$1,000
		28 Rent	$170

Balance the cash account at the end of the month. Extract the Trading and Profit and Loss accounts and finally extract the balance sheet.

15. (a) If the matrix $A = \begin{pmatrix} 2x & 8 \\ x & x^2 \end{pmatrix}$, prove that
$$|A^{-1}| = \frac{1}{2x(x-2)(x+2)}.$$

Find x for which the matrix A is a singular matrix.

(b) A transformation is represented by the matrix M where $M = \begin{pmatrix} 2 & 3 \\ 1 & 4 \end{pmatrix}$.
 (i) Find the image of point (-4,3) under M.
 (ii) Find the inverse of M.
 (iii) Given that point (7,9) is the image of the point (p,q) under M, find the value of p and q.

16. (a) A wire of 40 cm long is cut into two pieces. One of the pieces is $4x$ cm long. Each piece is then formed into a square. If y is the sum of the areas, prove that $y = 2x^2 - 20x + 100$. Draw the graph of $y = 2x^2 - 20x + 100$, hence state what the minimum value of area, y can be obtained.

(b) Two events A and B are independent. Given that $P(A \cap B) = \frac{1}{10}$ and $P(A) = \frac{1}{4}$. Find $P(A \cup B)$.

Examination Paper 15

TIME: 3 hours

Instructions

1. This paper consists of two sections A and B.
2. Answer **all** questions in section A and **any four** questions in section B.
3. All work done must be shown clearly.
4. Don't use electronic calculators for any question.
5. You are advised to spend no more than 2 hours in section A and using the remaining time in section B.

Section A

1. (a) The time is given as 2.613. Convert this in the form of hours, minutes, and seconds. Then round your answer to the nearest minutes.

   (b) Simplify $2.4 + \left(\dfrac{7.2}{1.6}\right) \times 4.6 - 13.1$

   (c) Find the smallest number that can be divided by 6, 12 and 15. Hence arrange the following fractions in ascending order.
   $$\frac{5}{6}, \frac{7}{12}, \frac{8}{15}$$

2. (a) If $\dfrac{\log a}{b} = \dfrac{\log b}{a}$, show that $\dfrac{a^a}{b^b} = 1$.

   (b) Given that $\sqrt[5]{64} = 2.297$, find $\sqrt[5]{2}$.

   (c) Simplify the expression $\dfrac{\sqrt{625x^3}}{\sqrt[3]{125x}}$.

3. (a) Express the function $f(x) = 2x^2 + 6x + 5$ in the form of
   $$a(x-2)^2 + b(x-2) + c$$
   where a, b and c are real numbers. Hence show that, when x is any value which is very close to 2, $f(x) \approx 14x - 3$. Find $f(2.1)$ correct to two significant figure.

   (b) Two finite sets A and B are non-disjoint sets with μ as their universal set. Draw a Venn diagram to represent these sets and shade the region which represent a set A∩(A'∪B). What is the simplest set that can represent the region you shade?

4. The lines l_1, l_2 and l_3 are given by equations

   l_1: $2x+y = 9$, l_2: $4x+3y = 23$ and l_3: $3x+2y = 15$

   Point A is the intersection of lines l_1 and l_2 and point B is the intersection of lines l_2 and l_3. Another variable point $C(x,y)$ lies on the same plane such that $BC = 2AC$,

   (a) Show that $3x^2+3y^2 - 18x - 22y+34 = 0$
   (b) Find y when $x = 5$.

5. (a) The exterior angles of a certain quadrilateral are a, b, c and d and their respective interior angles are e, f, g and h. Use an appropriate formula to show that $e+f+g+h = 360°$. Deduce that $a+b+c+d = 360°$.
   (b) On the triangle PQR, the angles at P and Q are 70° and 65° respectively. If N is the foot of the perpendicular from Q to PR, prove that $RN = QN$.

6. (a) According to Newton's law of gravitation, *the force that the two bodies attract one another is proportional to the product of their masses and inversely proportional to the square of their distance apart*. It is observed that, when two planets of masses 3×10^{20} kg and 5×10^{21} kg which are 5×10^6 km apart attract, their force of attraction is 4×10^{18} N. Find the force of attraction if two planets of masses 4×10^{22} kg and 3.5×10^{20} kg which are 8×10^8 km attract.
   (b) The radius of a circle varies as A^n where A is its area. What is the value of n?

7. (a) A man bought twenty hens at shs.10,000 each. During a year, he obtained from them 4,500 eggs which he sold at shs.4,000 per score. The cost of feeding them for the year was shs.300,000. At the end of the year he sold fourteen surviving hens at shs.5,000 each. Find his percentage profit. (1 score = 20 eggs)
   (b) The price of an article rose by 20% in the previous month and fell by 20% in this month, what is the total rise or fall percent?

8. (a) The sum of the first $3n$ terms of an arithmetic progression of the first term a, is twice the sum of the next n terms. Find the sum of the first $2n$ terms.
   (b) A certain kind of bacteria always splits itself into two in each second. If 20 bacteria were taken as a sample, how many bacteria will be in the sample after half a minute?

9. (a) Solve the equation $x^2 + 2x - 1 = 0$ by completing the square. Hence, or otherwise, show that
$$\tan 22\tfrac{1}{2}° = -1 + \sqrt{2}$$
given that $\tan 45° = 1$.

(b) Show that $\cos\theta + \sqrt{3}\sin\theta = 2\sin(30° + \theta)$. Hence solve the equation $\dfrac{1}{2}\cos\theta + \dfrac{\sqrt{3}}{2}\sin\theta - \cos 50° = 0$, if θ is an acute angle.

10. (a) Find the value of x from the triangle below.

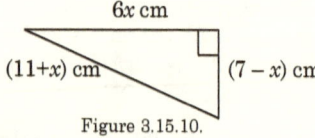

Figure 3.15.10.

(b) Use the method of substitution to solve for x and y from simultaneous equations
$$\begin{cases} \dfrac{x+5}{3} + \dfrac{y-2}{6} = \dfrac{3x-2}{5} - \dfrac{8-2y}{4} = 4 \end{cases}$$

Section B

11. A farmer alone can plant ten coconut trees and fifteen mango trees while his son can plant twelve coconut trees and ten mango trees per day. He buys each coconut tree he plants for shs.5,500 and each mango tree for shs.6,000. After growing, each coconut tree produce 40 coconuts and each mango tree produce 500 mangoes per period. How many days should each one plant the trees so as to produce at least 7,840 coconuts and 115,000 mangoes per period at a minimum cost?

12. The distribution of points for 50 students in a certain activity was recorded on the table below.

Class interval	10–14	15–19	20–24	25–29	30–34
Frequency	3	x	18		8

Given that the mean point is 23.3,

(a) Find the values of x and y.

(b) Use the formula to estimate the median point.

(c) Draw a histogram to represent the data and use it to estimate the mode point.

13. (a) A plane, 300 m above the ground level is observed from two points which are x metres apart. From one point which is due west of the plane, the angle of elevation is 30° and from the other point which is due south of the plane, the angle of elevation is 45°. Find the value of x.

(b) It is said that, 1 g of water has a volume of 1 cm³. In 3 hours, a circular pipe of diameter 4 cm was used to fill a tank which can hold 100 kg of water. Find the average speed of water in the pipe.

14. (a) The following are the information of transactions from Mr. Mazingo accounts. They were all paid in cash.

Capital	$10,500
Purchases	$3,000
Opening stock	$1,200
Sales	$8,200
Debtors	$2,900
Creditors	$560
Wages	$300
Furniture	$1,320
Rent	$340

Extract the trial balance for the transactions.

(b) From the information given in (a) above, find the closing stock and net profit if the total assets is $14,640. Hence determine the gross profit.

15. (a) A dealer has three deposits D_1, D_2 and D_3 keeping two types of machines M_1 and M_2. The number of machines in each deposit are shown on the table below.

	M_1	M_2
D_1	2	3
D_2	6	0
D_3	4	3

The cost of each machine is shown on the table below.

M_1	40€
M_2	30€

(i) Write the first table in the matrix form as matrix A with types of machines as columns and the second table as matrix B with costs as columns.

(ii) Find the product A×B

(iii) State the total cost that a dealer gets in each deposit.

(b) Given that the matrix $A = \begin{pmatrix} -1 & 2 \\ 2 & -1 \end{pmatrix}$ and $f(x) = x^2 + 2x$, find f(A). Hence find A^{-1} and deduce $|A|$.

16. (a) The relation R is defined as $R = \{(x,y): x^2 - y^2 = 4\}$
   (i) Find the domain and range of R.
   (ii) Find the x-intercepts and y-intercepts of R when possible.

(b) A bag contains two pair of gloves of different sizes. If two gloves are drawn at random, find the probability that the gloves drawn are of
   (i) the same size.
   (ii) the same hands.
   (iii) the matched pair.

Answers for Examination Papers 1-15

Examination Paper 1

1. (a) 5×10^{-2} (b) Given $\dfrac{403}{198}$

2. (a) $7 + 4\sqrt{2}$, 12.656
   (b) $x > -\dfrac{1}{2}$ or $x < -\dfrac{9}{2}$

3. (a) (i) {All integers}
       (ii) {All natural numbers}
   (b) £80,000

4. (a) (i) $\sqrt{317}$ (ii) $321°50'$
   (b) (i) 1:50000 (ii) 1.875 km^2

5. (a) (i) $\dfrac{29}{4}$ (ii) $\dfrac{3}{5}$
   (b) $x = \dfrac{-1 \pm \sqrt{43}}{7}$

6. (a) $y = -x + 9$
   (b) $x = 3$ and $y = 0$

7. (a) $k = 10$
   (b) $a = 10\sqrt{b}$, $\log a = 2\dfrac{1}{2}$

8. (a) 1,340
   (b) (i) 6 and $\dfrac{3}{2}$ respectively
       (ii) $0.3(6^{10} - 1)$

9. (a) 327 cm^2, 76.2 cm (b) $50°$

10. (a) 6 cm

11. 4 kg and 8 kg of food X and Y respectively

12. (b) 33.3 (c) 30–40, 32.9
    (d) 30–40, 33.5

13. (a) 4158 cm^2, 19.404 litres
    (b) 50

14. Dr   Cash A/c   Cr

Date	Particulars	Folio	Amount ($)	Date	Particulars	folio	Amount ($)
Jan,1	Capital		600,000	Jan,2	Purchases		150,000
3	Sales		100,000	5	Purchases		50,000
13	Sales		200,000	10	Trans. charge		5,000
22	Sales		320,000	18	Purchases		85,000
				30	Wages		3,500
				31	Balance	c/d	926,500
			1,220,000				1,220,000
Feb,1	Balance	b/d	926,500				

TRIAL BALANCE

Accounts	Dr	Cr
Capital	-	600,000
Purchases	285,000	-
Sales	-	620,000
Cash	926,500	-
Transport charges	5,000	-
Wages	3,500	-
	1,220,000	1,220,000

Dr	TRADING, PROFIT & LOSS ACCOUNTS		Cr
Purchases	285,000	Sales	620,000
Less: Closing stock	9,500		
Cost of sales	275,500		
Gross profit c/d	344,500		
	620,000		620,000
Transport charge	5,000	Gross profit b/d	344,500
Wages	3,500		
Net profit	336,000		
	344,500		344,500

BALANCE SHEET

Capital	600,000	Current assets	
Add: Net profit	336,000	Cash	926,500
		Stock	9,500
	936,000		936,000

15. (a) $-2i+3j$, $\sqrt{13}$

    (b) $a = 2$ and $b = \frac{1}{2}$

    or $a = -3$ and $b = -\frac{1}{3}$

16. (a) (ii) Domain = $\{x: x \neq 0\}$
    Range = $\{y: y > 0\}$

    (iii) π, no value

    (iv) No

    (b) (i) $\frac{2}{9}$ (ii) $\frac{1}{45}$ (iii) $\frac{31}{45}$

Examination Paper 2

1. (a) 4.983×10^3 (b) $5,040$

2. (a) £32,805 (b) $P = \sqrt[3]{\dfrac{R}{Q^3 - 2}}$

3. (a)

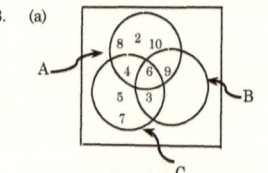

   7 and 1 respectively.
   (b) $4,000, $6,000 and $8,000 respectively.

4. (a) (i) $\sqrt{394}$ (ii) $x = -5, y = 11$
   (b) 48.4 N

5. (a) 56,500 (b) $x = 4, y = 6$

6. (a) $y = 2x+1$ (b) $(2,5)$
7. (a) £580.65 (b) ¥16483.3
   (c) $5391.3
8. (a) 3,708 (b) 400
9. (a) 18.84 cm^2

   (b) $R = \{(x,y): y \geq 0, y \leq -\frac{3}{2}x + 3,$

   $y \leq \frac{3}{4}x + 3\}$

10. (a) $x = 2$ cm (b) 2.82 cm
    (c) 11.71 cm (d) 12.04 cm
11. (a) (3,5) and (4,3) respectively

    (b) $x = \dfrac{-1 \pm \sqrt{5}}{2}$

12. (a) $2,563 (b) $2,343
    (c) (2500–2900), $2,492
13. (a) 8 cm (b) 10.8 cm
    (c) 53°07' (d) 48°

14.

Dr					Cash A/c		Cr
Date	Particulars	Folio	Amount (£)	Date	Particulars	folio	Amount (£)
Apr. 1	Capital		35,000	Apr 2	Office exp.		7,500
5	Sales		15,000	4	Purchases		8,000
18	Sales		18,000	8	Carriage		8,500
23	Sales		28,000	12	Purchases		10,000
				13	Garage exp.		6,000
				21	Purchases		6,000
				25	Wages		1,500
				30	Balance	c/d	48,500
			96,000				96,000
May.1	Balance	b/d	48,500				

TRIAL BALANCE

Accounts	Dr (£)	Cr (£)
Capital	-	35,000
Sales	-	61,000
Cash	48,500	-
Purchases	24,000	-
Office expenses	7,500	-
Carriage	8,500	-
Garage expenses	6,000	-
Wages	1,500	-
	96,000	96,000

15. (a) (i) $\begin{pmatrix} 2 & 3 \\ 3 & -4 \end{pmatrix}$ (ii) (23,-8)
    (b) $x = 3, y = 1$

16. (a) $a = 3$
    (b) (i) $2x^2 - 7x - 15$
        (ii) $3x - 2$ (iii) $2x - 2$
    (c) (i) 0.1   (ii) 0.75

Examination Paper 3

1. (a) $\dfrac{80}{101}$
   (b) 120 or 360 or 840 or 2520

2. (a) 1,238   (b) $\dfrac{3}{(x+1)(x+4)}$

3. (a) (i) {2,4,6,8,12,24}
       (ii) {Multiples of 6}

   (b)

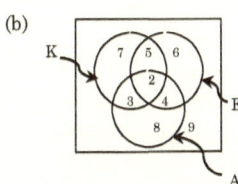

   K represents Kiswahili
   E represents English
   A represents Arabic
   (i) 7   (ii) 6
   (iii) 29   (iv) 9

4. (a) $\underline{r} = 7i - 11j$ (b) $\sqrt{170}$
   (c) 302°28'

5. (a) $a = 15°$
   (b) $\theta = 60°$ or $180°$ or $300°$

6. (a) 156 cm²   (b) 3.09 cm

7. (a) $y = -2x + 3$
   (b) $a = 13$ or $a = -2$, $(\dfrac{14}{5}, -\dfrac{13}{5})$

8. (a) (i) $P = \dfrac{13}{4} + \dfrac{3Q^2}{400}$ (ii) 3.73

9. (a) $\theta = 20°$

10. (a) (i) 1:20,000   (ii) 104cm
    (b) 4,092

11. 15 windows and 10 doors, shs.1,675,000

12. (a) (i) 3.14   (ii) 3   (iii) 1
    (b) (i) $\dfrac{47}{112}$ (ii) $\dfrac{1}{14}$

13. (a) (i) 42 hours   (ii) (06°N, 32°E)
    (b) The wind is moving from 052°05' at 199 knots

14. (a) $14,000   (b) $7,000
    (c) $25,000   (d) $16,000
    (e) $15,500

15. (a) (i) $\begin{pmatrix} 2 & 5 \\ 1 & 8 \end{pmatrix}$ (ii) 99 square units
    (b) $\begin{pmatrix} 12 & 8 \\ 6 & 28 \end{pmatrix}$

16. (a) (i) Domain = {$x: x \in \Re$}
            Range = {$y: y > 4$}
        (ii) $f^{-1}(x) = \log_2(x-4)$
    (b) (i) $\dfrac{17}{40}$ (ii) $\dfrac{23}{40}$

Examination Paper 4

1. (a) 1.709 (b) (i) 60,000 (ii) 50%
2. (a) $x = 50°, y = 40°$
   (b) $\dfrac{39 + 32\sqrt{2}}{10}$
3. (a) (i)

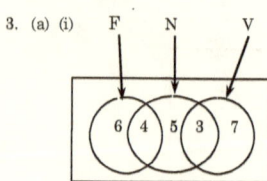

   F Represents football
   V represent volleyball
   N represents Netball
   (ii) 8   (iii) 33

   (b) $x \geq \dfrac{5}{2}$

4. (a) (i) $\begin{pmatrix} 5 \\ -1 \end{pmatrix}$ and $\begin{pmatrix} 2 \\ -2 \end{pmatrix}$ respectively
   (ii) $\sqrt{58}$
   (b) 517.3 km
5. (a) 61 kg
6. (a) (i) $x = \dfrac{3y}{z}$ (ii) 1.8
   (b) 8.4%
7. (a) (i) (6,-5), (5,-4) respectively
   (ii) $x - y = 9$
   (b) 7.6 units
8. (a) 15 terms (b) 4 and 16
9. (a) (i) $\dfrac{642}{289}$ (ii) $-\dfrac{17}{7}$
10. (a) 10 cm
    (b) 8.2 cm and 3.08 cm respectively
    (c) 6.75 cm²
11. 15 televisions and 20 radios
12. (a) (i) 5 (ii) 5 (iii) 5
    (b) (i) $x = 4°, y = 5°$ (ii) shs.27,300
13. (a) 12.45 cm (b) 62°48′
    (c) 68°07′ (d) 132.4 cm³

14.

Dr   Cash A/c   **Cr**

Date	Particulars	Folio	Amount (£)	Date	Particulars	folio	Amount (£)
Sep, 1	Capital		100,000	Sep, 2	Purchases		55,000
15	Sales		82,000	10	Purchases		40,000
23	Madevu		15,000	30	balance	c/d	102,000
			197,000				197,000
Oct, 1	Balance	b/d	102,000				

Dr   Capital A/c   **Cr**

Date	Particulars	Folio	Amount (£)	Date	Particulars	folio	Amount (£)
Sep, 30	Balance	c/d	100,000	Sep, 1	Cash		100,000
			100,000				100,000
				Oct, 1	balance	b/d	100,000

Dr   Purchases A/c   **Cr**

Date	Particulars	Folio	Amount (£)	Date	Particulars	folio	Amount (£)
Sep, 2	Cash		55,000	Sep, 30	balance	c/d	95,000
10	Cash		40,000				
			95,000				95,000
Oct, 1	Balance	b/d	95,000				

Dr   Sales A/c   **Cr**

Date	Particulars	Folio	Amount (£)	Date	Particulars	folio	Amount (£)
Sep, 30	Balance	c/d	107,000	Sep, 5	Madevu		25,000
				15	Cash		82,000
			107,000				107,000
				Oct, 1	balance	b/d	

Dr			Madevu A/c				Cr
Date	Particulars	Folio	Amount (£)	Date	Particulars	folio	Amount (£)
Sep, 5	Cash		25,000	Sep, 23	Sales		15,000
				30	balance	c/d	10,000
			25,000				25,000
Oct, 1	Balance	b/d	10,000				

15. (a) (i) 7 terms

   (ii) $\dfrac{463}{64}\begin{pmatrix}\dfrac{12}{5} & -2 \\ 4 & -\dfrac{9}{5}\end{pmatrix}$

   (b) (i) (19,12)   (ii) $(-\dfrac{17}{7},\dfrac{23}{14})$

16. (a) (i) $(\dfrac{2}{3},\dfrac{17}{3})$ (ii) $x=\dfrac{2}{3}$

   (iii) Minimum value, $y=\dfrac{17}{3}$

   (b) (i) 0.5   (ii) 0.6

   (c) (i) $\dfrac{9}{40}$ (ii) $\dfrac{5}{8}$

Examination Paper 5

1. (a) 0.02547   (b) 2.5×10^{-2}

2. (a) $-\dfrac{1}{2}$

   (b) $\dfrac{2^{100}}{(2\times5)^{50}}=\dfrac{2^{50}}{5^{50}}=\left[\left(\dfrac{2}{5}\right)^2\right]^{25}$

   $=0.16^{25}$

   (c) $(a-b)(a-b-1)(a-b+1)$

3. (a) (i) 4   (ii) 5

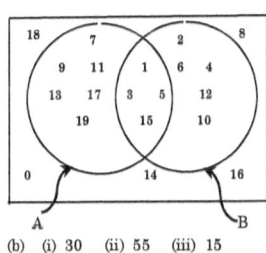

   (b) (i) 30   (ii) 55   (iii) 15

4. (a) $a=3, b=2$ (b) $25i+48j$
   (c) 10
5. (a) (i) $x=3y$ (ii) 24.6
   (b) (i) 50.5m   (ii) 40 seconds
6. (a) (i) $T(3.5, 2.5)$ and $R(0.5, 1.5)$
   (ii) 5.7 units
   (b) (i) $\dfrac{7}{5}$ and -2
   (ii) $10x^2+7x+10=0$
7. (a) 8.7 cm or 4.1 cm
8. (a) ± 3 (b) 10 years
9. (a) (i) 35°   (ii) 55°
   (b) (53°20′S, 03°29′E)
10. (a) 519.75 cm³
11. (a) 21 at point of coordinates $x=2, y=5$
12. (b) 45.7   (c) 41.9
13. (a) (i) 18   (ii) -15
   (b) (i) 8 cm and 6 cm
   (ii) 10.77 cm

14. Dr   TRADING PROFIT AND LOSS A/C   Cr

Opening stock	3,250	Sales	82,000
Purchases	48,000		
Cost of goods available	51,250		
Less: Closing stock	4,800		
Cost of goods sold	46,450		
Gross profit c/d	35,550		
	82,000		82,000
Insurance	5,470	Gross profit b/d	35,550
Wages	6,580		
Bad debts	6,200		
Net profit	17,300		
	35,550		35,550

BALANCE SHEET

Capital	40,000		Fixed assets	
Add: Net profit	17,300		Land and buildings	10,000
	57,300		Furniture	9,800
Less: Drawing	4,000		Current assets	
		53,300	Cash	30,500
Long term liabilities			Stock	4,800
Loan from bank		6,000	Debtors	8,800
Short term liabilities				
Creditors		4,600		
		63,900		63,900

15. (a) (i) $y = \frac{1}{3}x + 4$ (ii) (-3.4, 5.2)

    (b) $\begin{pmatrix} 9 & 8 \\ 16 & 17 \end{pmatrix}$ and $\begin{pmatrix} 4 & 0 \\ -5 & 9 \end{pmatrix}$

16. (a) $f(x) = -2(x+1)^2 + 8$

    (b) Range = $\{y: -\frac{49}{8} \leq y \leq 247\}$

    (c) (i) 0.8099 (ii) 0.9901
    (iii) 0.099

Examination Paper 6

1. (a) 50 cm³
   (b) 1.5×10^{-1}
2. (a) $a = 1.228 \times 10^{-11}$
   (b) a = 50 or 0.05
3. (a) 16.5 cm and 27.5 cm
   (b) (i) 6,160 cm³ (ii) 11.4 cm
4. (a) (i) -1 (ii) $x = 1$
   (b) $2x+4$, 13

5. (a)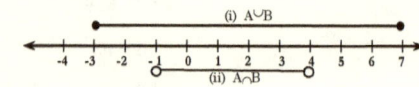

   (b)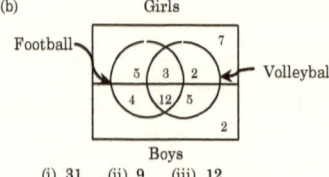
   (i) 31 (ii) 9 (iii) 12

6. (a) (i) 10 units (ii) 10 units
   (b) (i) $-\underline{x} - \frac{1}{3}\underline{y}$ (ii) $\frac{1}{6}\underline{y} - \frac{1}{2}\underline{x}$
   (iii) $\frac{1}{2}(\underline{x} + \underline{y})$

7. (a) $x+y = 2$ and $y - x = 2$ respectively
   (b) B = (2,4), D = (3,5)
   (c) $(2\sqrt{10} + 6\sqrt{2})$ units
   or 14.81 units

8. (a) 23 cm
   (b) (i) 15.89 cm (ii) 38°25′
   (iii) 98.74 cm²
9. (a) 5, 9 and 13, $4n+1$
   (b) 3 and 4 or -3 and -8
10. (a) △ABC ≡ △CDA
    (b) △XEA, △ABC, △CDA
    (c) EX = 3.2 cm, XF = 4.8 cm
11. (a) $x = 30°$, 90° and 150°
    (b) $x = 1$ and $y = 2$
12. (a) $x = 7, y = 5$ (b) 50 (c) 52
13. (a) 327 cm²
    (b) (i) 113.0 cm³ (ii) 410.3 cm³

14. Dr   TRADING PROFIT AND LOSS A/C   Cr

Opening stock	8,700	Sales	125,550
Purchases	94,300		
Cost of goods available	103,000		
Less: Closing stock	8,000		
Cost of goods sold	95,000		
Gross profit c/d	30,550		
	125,550		125,550
Wages	14,000	Gross profit b/d	30,550
Rent	11,000		
Net profit	5,550		
	35,550		30,550

BALANCE SHEET

ASSETS		Capital	75,000
Cash	72,550	Add: Net profit	5,550
Opening stock	8,000		
	30,550		30,550

15. (a) (i) $\begin{pmatrix} 1 & 1 \\ 2 & -1 \end{pmatrix}$ (ii) $y = -3$

     (b) $a = \pm 3$, $b = 1$, $(-1,-2)$

     (c) (i) $\begin{pmatrix} 35 & 10 \\ -10 & 15 \end{pmatrix}$ (ii) $\begin{pmatrix} \frac{4}{11} & \frac{15}{22} \\ -\frac{14}{33} & \frac{19}{66} \end{pmatrix}$

16. (a) (i) Domain = $\{x: x \in \Re\}$,
           Range = $\{y: y > 0\}$
       (ii) $x = -4$

     (b) (i) 3.32 million
       (ii) 27.4 years

     (c) (i) 144   (ii) 54   (iii) $\frac{1}{2}$

Examination Paper 7

1. (a) (i) 4.86×10^{-2} (ii) 1.5×10^0

    (b) $\frac{161}{36}$

2. (b) (i) $\overline{1}.5562$ (ii) 1.2219

    (c) $\frac{22}{7}$

3. (a)

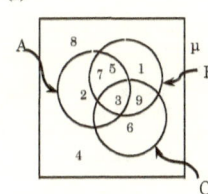

     (i) $\{3,5,7\}$ (ii) $\{1,3,5,7,9\}$

     (iii) $\{1,2,3,5,6,7,9\}$

    (b) $(-2,2)$

4. (a) shs.65,600

    (b) (i) $a = 2b^2 + \frac{3}{b}$ (ii) 72.5

5. (a) 24.6 N in the direction of 146°07′ from positive x-axis

    (b) $x = 2, y = 3, z = 4$

6. (a) $y = 15°$

    (b) 29.13 cm², 20.57 cm

7. (a) $a = 3$ or $a = -\frac{4}{3}$

    (b) 4 or -2.5

    (c) $y = 4x + 3$ or $y = -\frac{5}{2}x - \frac{4}{3}$

8. (a) (i) 3, 7   (ii) 7, 7

    (b) (i) $129,000

       (ii) After 15 months

9. (a) 120°
    (b) (i) 33.03 cm   (ii) 57.7 cm²

10. (a) (i) $\frac{\sqrt{6}+\sqrt{2}}{4}$ (ii) $-2 + \sqrt{3}$

     (b) (i) $\frac{1 - 2\sqrt{6}}{6}$ (ii) $\frac{\sqrt{3} - 2\sqrt{2}}{6}$

11. (a) 30 boxes of pencils and 10 boxes of pens
     (b) 56%
     (c) 20 boxes of pencils and 10 boxes of pens
     (d) shs.19,200

12. (a) 111.85
     (b) (i) 111.8   (ii) 33 students

13. (a) 10.22 cm   (b) 60°38′
     (c) 246.17 cm²

14. (b)

Accounts	Dr ($)	Cr ($)
Capital	-	80,000
Cash	118,600	-
Sales	-	134,900
Purchases	63,500	-
Furniture	25,000	-
Carriage	800	-
Wages	2,000	-
Debtors	5,000	-
	214,900	214,900

15. (a) (ii) 4
    (b) (i) (10,25) (ii) (-2,-13)
    (iii) $\begin{pmatrix} 2 & -\frac{8}{3} \\ 1 & 0 \end{pmatrix}$ (iv) $(\frac{14}{3}, -7)$

16. (b) (i) $\frac{13}{80}$ (ii) $\frac{11}{40}$

Examination Paper 8

1. (a) shs.873,750
   (b) (i) 8.36×10^{-3} (ii) 7
2. (a) $a = 5$ and $b = 5$
   (b) $-14 \leq x < -8$ or $-4 < x \leq 2$
3. (a) 124 (b) 12, 28
4. (a) (i) $\begin{pmatrix} -4 \\ 2 \end{pmatrix}, \begin{pmatrix} 2 \\ -6 \end{pmatrix}, \begin{pmatrix} -6 \\ 8 \end{pmatrix}$
   (ii) 10
   (b) (i) 195 km/h (ii) 23°20′
   (iii) 096°40′
5. (a) $A = \frac{44}{7} r(r+h)$ (b) 748
   (c) $x = 2, y = 2, A = 2\pi r(r+h)$
6. (a) $x = 2$ or $x = -4$
   (b) $x = 3$ or $x = -1$
7. (a) (i) 1:50,000 (ii) 300,000 m^3
   (b) 16 cm^2
8. (a) (i) $(\frac{10}{3}, -\frac{7}{3})$ (ii) $\frac{8\sqrt{5}}{3}$
   (b) $(x-3)^2 + (y+5)^2 = 7^2$
   (i) (3,-5), 7 units
   (ii) 154 square units
9. (a) (i) $\frac{4}{3}$ or $\frac{3}{4}$
   (ii) 9, 12 and 16 or 16, 12 and 9 respectively
   (iii) 64
   (b) 260, 4(n+1)
10. (a) 1,256 cm^2, 125.6 cm
    (b) 15 cm
11. (a) $a = 2, b = 1, c = 4$
    (b) {x+y ≤ 100, x+y ≥ 40, x ≤ 70, y ≤ 50, x ≥ 0 and y ≥ 0}
12. (a) 77 kg (b) (i) 79 kg (ii) 33
13. (a) 6.2 cm, 482.8 cm^2
    (b) 6.49 cm

14.

Dr   TRADING PROFIT AND LOSS A/C   Cr

Opening stock		978,780	
Purchases	6,560,000		
Less: R. outward	584,700		
Net Purchases		5,975,300	
Cost of goods available		6,954,080	
Less: Closing stock		860,000	
Cost of goods sold		6,094,080	
Gross profit c/d		3,130,100	
		9,224,180	
Salaries		778,000	
Wages		685,000	
Rent		136,300	
Insurance		1,200,000	
Net profit		330,800	
		3,130,100	

Sales	9,900,000
Less: R. inward	675,820
	9,224,180
Gross profit b/d	3,130,100
	3,130,100

Examination Papers

BALANCE SHEET

Capital		5,000,000		Fixed assets		
Add:Net profit		330,880		Fixture and fittings		800,000
		5,330,800		Motor vehicles		750,000
Less: Drawing		1,000,000		Current assets		
			4,330,800	Cash		1,500,000
Liabilities				Stock		820,800
Creditors			400,000	Debtors		860,000
			4,730,800			4,730,800

15. (a) (i) $\begin{pmatrix} \frac{1}{7} & -\frac{3}{7} \\ \frac{1}{7} & \frac{4}{7} \end{pmatrix}$ (ii) $\begin{pmatrix} -\frac{11}{43} & \frac{4}{43} \\ \frac{8}{43} & \frac{1}{43} \end{pmatrix}$

(b) 57.5 square units

(c) P'(-5,4), Q'(1,6), R'(5,2)

16. (a) (ii) Domain = $\{x: x > 3\}$, Range = $\{y: y \geq 0\}$

(iii) 2, no values of $f(-6)$

(b) (i) $\frac{2}{21}$ (ii) $\frac{9}{35}$ (iii) $\frac{2}{21}$

Examination Paper 9

1. (a) 3,780 (b) $\frac{53}{66}$
2. (a) $x = -1, y = 0.5$
   (b) (i) 1.6990 (ii) $\overline{1}.3980$
3. (a) (i) 7 (ii) 4 (b) 25.14
4. (a) $x = 3$ or $x = -\frac{19}{3}$
   (b) (i) 100° (ii) 10°
5. (a) 110°
   (b) 999×55 = 55(1000 − 1)
       = 55000 − 55 = 54945
   (c) $(x+2)(xy+y+1)$
6. (a) $V = 4190.48 cm^3$
   (b) $V = \frac{88}{21} r^3$
   (c) $a = \frac{4}{3}$, $V = \frac{4}{3}\pi r^3$
7. (a) (2,2) (b) 5 square units
8. (a) 1.5
   (b) 7 and 4 respectively
9. (a) $1,800, $1,200
   (b) 196 cm²
10. (a) 34.64 m, 23.09 m
    (b) AB = 6.19cm, ∠A = 81°53′, ∠C = 48°07′
11. 4 and 7 products of type A and B respectively.
12. (a) 49 (b) 45.6 (c) 46.29
13. (a) 20.54cm²
    (b) (i) 08:46 p.m. (ii) 7,653 km

14. (a)

Account	Dr ($)	Cr ($)
Capital	-	100,000
Cash	72,000	-
Opening stock	50,000	-
Purchases	161,580	-
Sales	-	223,580
Wages	24,000	-
Rent	10,000	-
Rates	6,000	-
	323,580	323,580

(b) $52,000 and $24,000 respectively

BALANCE SHEET

Capital	100,000	Current Assets	
Add: Net profit	24,000	Cash	72,000
		Stock	52,000
	124,000		124,000

15. (a) $\begin{pmatrix} \frac{5}{22} & \frac{3}{22} \\ \frac{2}{11} & -\frac{1}{11} \end{pmatrix}$ (b) $x = 1, y = -1$

    (c) (i) $\begin{pmatrix} 0 & 1 \\ 1 & 0 \end{pmatrix}$, reflection about line $y = x$
        (ii) $(4, -2)$

16. (a) $f(x) = -2(x+1)^2 + 10$
    (i) $(-1, 10)$
    (ii) $x = -1$
    (iii) Maximum value, $y = 10$
    (iv) Range = $\{y : y \leq 10\}$

    (b) $\frac{11}{12}$

    (c) (i) $\frac{11}{15}$
        (ii) Mutually exclusive events

Examination Paper 10

1. (a) (i) $a \approx 22.5$, $b \approx 52.0$, $c \approx 25.0$
       (ii) 46.8
   (b) $\frac{37}{66}$

2. (a) $x = 30$ (b) $n = 1.58$

3. (a) (i)

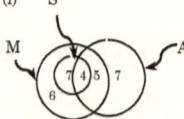

   M represents Mathematics
   S represents Science
   A represents Arts
       (ii) 12   (iii) 29
   (b) (i) $\frac{63}{65}$ (ii) $\frac{16}{65}$

4. (a) $2\sqrt{13}$ (b) $30\sqrt{7}$ N (c) $x = 2, y = 3$

5. (a) $80(2x+1)(2x^2+2x+13)$
   (b) $a = -2$, $b = -3$ and $c = -4$

6. (a) 907.5 square units
   (c) 106.8 units

7. (a) (i) $P = 2Q + \frac{4}{Q}$ (ii) 9
   (b) 5.45 hours

8. (a) 3, 8
   (b) 3 and 4 or -3 and -6 respectively

9. (a) 509.2 cm²   (b) 10°

10. (a) (i) 1:200,000 (ii) 80 km²
    (b) (i) 5 cm   (ii) 15 cm

11. (a) shs.27,000 and shs.16,000
    (b) 90 eggs

12. (a) $x = 13$ (b) 11.7
    (c) 10.09

13. (a) (i) 40°25′ (ii) 1,867cm³ (b) (i) 20,000km
    (ii) 166 hrs, 40 min

14. Dr TRADING AND PROFIT AND LOSS A/C Cr

Opening stock	9,500	Sales	100,000
Purchases	85,500		
Cost of goods available	95,000		
Less: Closing stock	5,000		
Cost of goods sold	90,000		
Gross profit c/d	10,000		
	100,000		100,000
Expenses	7,500	Gross profit b/d	10,000
Net profit	2,500		
	10,000		10,000

15. (a) (i) $\begin{pmatrix} -2 & 1 \\ 1.5 & -0.5 \end{pmatrix}$

    (ii) $\begin{pmatrix} 0 & 1 & 2 \\ 3 & 4 & 2 \end{pmatrix}$, 2×3

(b) (8,0)   (c) (18,37)

16. (a) (ii) Domain = $\{x: x \neq 0\}$
    Range = $\{y: y \neq 0\}$
   (iii) 9
   (iv) One-to-one function because each element of the range is corresponding to only one element of the domain

  (b) (i) 20   (ii) $\dfrac{13}{20}$

Examination Paper 11

1. (a) 23.4, 19.5, 52.0 (b) 168
2. (b) $x = -\dfrac{1}{5}$ (b) $\dfrac{5}{4}$
3. (a)

   (i) {a,b,c,d,e,f,g,h,i}
   (ii) {a,b,c,d,e,f,g}

(b) (i) 11 (ii) 16

4. (a) (i) $t = -\dfrac{2}{5}$ (ii) $2\sqrt{37}$

  (b) $\sqrt{26}, -\dfrac{1}{\sqrt{26}}$ and $\dfrac{5}{\sqrt{26}}$

5. (a) 4.8
6. (a) 9 (b) 15 days
7. (a) £4,000, £60,000 and £8,000 respectively
  (b) £4,200
8. (a) 7, 12, 17 (b) 2, 3
9. (a) 1080° (b) 9
10. (a) $x^2 + 2x = 3 \Rightarrow x^2 = 3 - 2x$
    $\Rightarrow x^4 = (3-2x)^2 = 9 - 12x + 4x^2$
    $= 9 - 12x + 4(3 - 2x)$
    $\Rightarrow x^5 = 21x - 20x^2$
    $= 21x - 20(3 - 2x)$
    $\Rightarrow x^5 = 61x - 60$
  (b) $-5 \leq x \leq 3$

11. (b) 26 at (2,5)
12. (a) 59.8 (b) 52 (c) 56.9
13. (a) 10,000 km (c) $10\sqrt{3}$ cm

14.

Dr   Cash A/c   Cr

Date	Particulars	Folio	Amount ($)	Date	Particulars	folio	Amount ($)
May, 1	Capital		80,000	May, 2	Purchases		35,000
9	Sales		30,000	16	Wages		6,000
				21	Asha		7,000
				25	Rent		4,500
				31	balance	c/d	57,500
			110,000				110,000
Jun, 1	Balance	b/d	57,500				

Dr   Sales A/c   Cr

Date	Particulars	Folio	Amount ($)	Date	Particulars	folio	Amount ($)
May, 31	balance	c/d	58,000	May, 5	Makame		28,000
				9	Cash		30,000
			58,000				58,000
				June, 1	Balance	b/d	58,000

Dr   Purchases A/c   Cr

Date	Particulars	Folio	Amount ($)	Date	Particulars	folio	Amount ($)
May, 2	Cash		35,000	May, 31	balance	c/d	60,000
13	Asha		25,000				
			60,000				60,000
June, 1	Balance	b/d	60,000				

Examination Papers

Dr			Makame's A/c				Cr
Date	Particulars	Folio	Amount ($)	Date	Particulars	folio	Amount ($)
May, 5	Sales		28,000	May, 31	Balance	c/d	28,000
			28,000				28,000
Oct, 1	Balance	b/d	28,000				

Dr			Asha's A/c				Cr
Date	Particulars	Folio	Amount ($)	Date	Particulars	folio	Amount ($)
May, 21	Cash		7,000	May, 13	Purchases		25,000
31	balance	c/d	18,000				
			25,000				25,000
				Jun, 1	Balance	b/d	18,000

TRIAL BALANCE AS ON 31st MAY

Account's name	Dr ($)	Cr ($)
Capital	-	80,000
Cash	57,500	-
Wages	6,000	-
Purchases	60,000	-
Sales	-	58,000
Creditors	-	18,000
Debtors	28,000	-
Rent	4,500	-
	156,000	156,000

15. (a) $\begin{pmatrix} -3 & 0 \\ 0 & -3 \end{pmatrix} \begin{pmatrix} 1.5 & -2 \\ -0.5 & 1 \end{pmatrix}$

16. (a) (i) $\frac{2}{3}$ (ii) $\frac{2}{5}, \frac{11}{15}$
    (b) $(-1, -7)$, $x = -1$,
    Range = $\{y: y \geq -7\}$

Examination Paper 12

1. (a) 0.1306
   (b) 16 years 3 months

2. (a) $-\frac{5}{3}$
   (b) 7.706

3. (a)

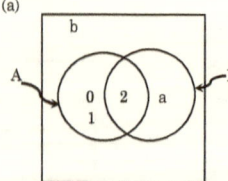

   (i) $\{0,1\}$ (ii) $\{0,1,a,b\}$
   (b) $n = \frac{lR}{Q - lP}$

4. (a) **b** − **a**, $k = 2$ (b) $k = \pm 2$

5. (a) $\frac{1}{2}ab$
   (b) (i) 6 cm (ii) 1.5×10^{-3} km²

6. (a) 3 men
   (b) $y \propto \frac{\sqrt{x}}{\sqrt{z}}$, $y^2 = \frac{36x}{z}$

7. (a) $6\frac{2}{3}\%$
   (b) (i) 3 years (ii) $5,000

8. (a) 1,425 (b) 3, 4

9. (a) $x = 15°$ (b) 6.43 cm

10. (a) $(\frac{px - qy}{p + q})\%$
    (b) $A = 1, B = 2, \frac{13}{42}$

11. 125 first class and 375 economy class tickets

12. (a)

Age (years)	Frequency
10	2
11	4
12	5
13	5
14	2
15	2

Mode = 12 or 13, Median = 12

(b) 12.35 years (c) 15.35 years

13. (a) 2352 cm³ (b) 1528 cm²

Examination Papers

14. (a)

Dr		TRADING ACCOUNT		Cr
Opening stock	6,000		Sales	50,000
Purchases	35,000		Less: Ret. in	1,000
Less: Ret. out	3,500			49,000
Net Purchases		31,500		
Cost of goods available		37,500		
Less: Closing stock		5,800		
Cost of goods sold		31,700		
Gross profit c/d		17,300		
		49,000		49,000

Dr		PROFIT AND LOSS ACCOUNT		Cr
Expenses	25,000		Gross profit b/d	17,300
			Net Loss	7,700
	25,000			25,000

∴ The gross profit is £17,300/= and Net Loss is £7,700/=

15. (a) $x = 3, y = 2$
    (b) $(1,3)$
    (c) $(-2\sqrt{3}-1, \sqrt{3}-2)$

16. (a) $A = l(250 - l)$
    (b) $A = -(l - 125)^2 + 15625$
    (c) $l = 125$,
    Maximum area = 15625m^2

Examination Paper 13

1. (a) 1.0
   (b) (i) $\dfrac{5}{21}$ (ii) 21 dollars

2. (a) $x = 3$ (b) $\dfrac{-1+2\sqrt{2}}{7}$

3. (a) (i) 10% (ii) 15%
   (b) $1+6x+12x^2+8x^3$, 2,744

4. (b) $\dfrac{5}{13}, \dfrac{12}{13}$, Direction cosines
   (c) $\dfrac{1}{13}(5\mathbf{i}+12\mathbf{j})$

5. (a) Angles N = 54°, B = 41° and lines LN = 1.3 cm, PB= 3.6 cm
   (b) $x = 10°, y = 40°$

6. (a) 4.5 km
   (b) (i) inversely (ii) $yx = 25$

7. (a) 5:8:15:16 (b) 20%

8. (a) 20, 21 terms (b) $\dfrac{11}{2}$

9. (b) 5 cm, $2\sqrt{14}$ cm

10. (a) $x = \pm 2$ and $y = \pm 3$
    or $x = \pm \dfrac{\sqrt{61}}{61}$ and $\pm \dfrac{42\sqrt{61}}{61}$

11. Rides 12 km on 30 km/h, and 24 km on 40 km/h

12. (a) 46.8 (b) 30 (c) 78

13. (a) 140° (b) 9.33 cm, 76.64 cm³

14. (a) $12,000 (b) $16,000
    (c) $4,000 (d) $1,000
    (e) $34,000 (f) $2,000

15. (a) (i) $\begin{pmatrix} -2 & 7 \\ -4 & 11 \end{pmatrix}$
    (b) (i) $\dfrac{1}{9}\begin{pmatrix} 5 & -3 \\ -7 & 6 \end{pmatrix}$

16. (a) (i) $\{x: x \geq 1\}$
    (ii) $f^{-1}(x) = 1 + \sqrt{x+5}$
    (b) The graph represents a function because each element of the domain is corresponding to only one element of the range.

Examination Paper 14

1. (a) $\dfrac{1}{4}$, 500 g (b) 0.142

3. (b) (i) $\{1,2,3,4,5,6\}$ (ii) $\{6,5\}$
   (iii) $\{1,2,5\}$

4. (a) $\mathbf{a}-\mathbf{b}, \dfrac{2}{5}\mathbf{a}-\mathbf{b}, p = -\dfrac{3}{2}, q = \dfrac{5}{2}$
   (b) F = 15 N or F = 60 N

5. (a) $48\sqrt{3}$ cm² (b) 237.8 cm²

6. (a) RPS 488 (b) 200 years

7. (a) £2,000, £4,000, £8,000
   (b) 19%

8. (b) $|x| > 1, \dfrac{x^2}{x^2-1}, 1$

9. (b) $\dfrac{a}{c}, \dfrac{b}{c}$

10. (a) Decagon and Pentagon respectively.
    (b) $x = 2, y = 3$
11. 60 kg of F_1 and 80 kg of F_2.
12. (a) 33.3 (b) 31.9
13. (a) 3,889 km (b) 20°

14.

Dr   Cash A/c   Cr

Date	Particulars	Folio	Amount ($)	Date	Particulars	folio	Amount ($)
Sept, 1	Capital		4,500	Sept, 3	Purchases		1,500
5	Sales		2,500	4	Wages		350
11	Sales		1,800	13	Insurance		110
				15	Purchases		1,150
				20	Drawings		1,000
				28	Rent		170
				30	balance	c/d	4,520
			8,800				8,800
Jun, 1	Balance	b/d	4,520				

Dr   TRADING AND PROFIT AND LOSS A/C   Cr

Purchases	2,650	Sales	4,300
Gross profit c/d	1,650		
	4,300		4,300
Wages	350	Gross profit b/d	1,650
Insurance	110		
Rent	170		
Net profit	1,020		
	1,650		1,650

BALANCE SHEET

Capital	4,500	Current Assets	
Add: Net profit	1,020	Cash	4,520
	5,520		
Less: Drawings	1,000		
	4,520		4,520

15. (a) $x = 0$ or $x = -2$ or $x = 2$
    (b) (i) (1,8) (ii) $\dfrac{1}{5}\begin{pmatrix} 4 & -3 \\ -1 & 2 \end{pmatrix}$
    (iii) $p = \dfrac{1}{5}, q = \dfrac{11}{5}$

16. (a) 50 cm² (b) $\dfrac{11}{20}$

Examination Paper 15

1. (a) 2:36:47, 2:37 (b) 10
   (c) 60, $\dfrac{8}{15}, \dfrac{7}{12}, \dfrac{5}{6}$
2. (b) 1.149 (c) $5x\sqrt[6]{x}$
3. (a) $f(x) = 2(x-2)^2 + 14(x-2) + 25$, 26
   (b)

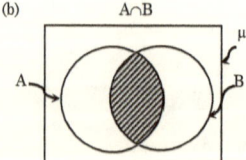

A∩B

4. (b) $y = 1$ or $y = \dfrac{19}{3}$
6. (a) 1.46×10^5 N (b) $n = 0.5$
7. (a) 94% (b) 2% fall
8. (a) $\dfrac{14an^2}{5n+1}$ (b) $5(2)^{31}$ bacteria
9. (a) $x = -1 \pm \sqrt{2}$ (b) $\theta = 10°$
10. (a) $x = 2$ (b) $x = 4, y = 8$
11. 10 days by a farmer, 8 days by his son.
12. (a) $x = 9, y = 12$ (b) 23.11
    (c) 22.5
13. (a) 600 m (b) 26.54 m/h

14. (a)

Account's name	DR ($)	CR ($)
Capital	-	10,500
Purchases	3,000	-
Opening stock	1,200	-
Sales	-	8,200
Debtors	2,900	-
Creditors	-	560
Wages	300	-
Furniture	1,320	-
Rent	340	-
Cash	10,200	-
	19,260	19,260

(b) $220, $3,580

15. (a) (i) $A = \begin{pmatrix} 2 & 3 \\ 6 & 0 \\ 4 & 3 \end{pmatrix}$, $B = \begin{pmatrix} 40 \\ 30 \end{pmatrix}$

   (ii) $\begin{pmatrix} 170 \\ 240 \\ 250 \end{pmatrix}$

   (iii) 170€, 240€ and 250€ respectively

(b) $\begin{pmatrix} 3 & 0 \\ 0 & 3 \end{pmatrix}, \frac{1}{3}\begin{pmatrix} 1 & 1 \\ 2 & 2 \end{pmatrix}, -3$

16. (a) (i) Domain = {$x: x \leq -2$ or $x \geq 2$}

   Range = {$y: y \in \Re$}

   (ii) x-intercept = ± 2, no y-intercept

(b) (i) $\frac{1}{3}$ (ii) $\frac{1}{3}$ (iii) $\frac{1}{3}$

www.ingramcontent.com/pod-product-compliance
Lightning Source LLC
Chambersburg PA
CBHW030922180526
45163CB00002B/435